AF852782

Changing Lanes

A Transition into Critical Infrastructure Cyber Defense

William Dove, Ph.D., CISSP

First Edition

Changing Lanes: A Transition into Critical Infrastructure Cyber Defense

For permission requests, contact: will@dovecybergrp.com

First edition, 2026

ISBN: 979-8-9989961-5-3

Cover design by: William Dove

Printed in the United States of America

Disclaimer

This book is intended for educational and informational purposes only. It is designed to support professional understanding of incident response, critical infrastructure cyber defense, operational decision-making, and resilience in complex environments. It is not intended to provide legal advice, engineering certification, regulatory approval, or operational authorization for any specific action, system, or organization.

Readers are responsible for applying independent professional judgment and for complying with all applicable laws, regulations, contracts, policies, safety requirements, and organizational procedures. Because cybersecurity and critical infrastructure environments vary widely in architecture, mission, legal constraints, and operational consequences, no single approach is appropriate in every situation. Actions discussed in this book should be evaluated in light of the reader's specific environment, authorities, and responsibilities.

The author makes no guarantee that the methods, concepts, or practices discussed in this book will prevent incidents, eliminate risk, or ensure any particular outcome.

Cybersecurity, operational technology, emergency management, and critical infrastructure defense are dynamic fields, and conditions, threats, standards, and technologies continue to evolve.

Any references to organizations, standards, technologies, incidents, sectors, or operational practices are provided for educational discussion only and do not constitute endorsement unless expressly stated. The views expressed in this book are those of the author and do not necessarily reflect the official position of any employer, agency, military organization, public institution, or affiliated entity.

Readers should consult appropriate legal, technical, engineering, operational, and regulatory professionals before applying concepts from this book in real-world environments.

Preface

This book was written for people stepping into one of the most demanding areas of cybersecurity: the defense of critical infrastructure. It is intended for military cyber personnel transitioning into civilian critical infrastructure defense roles, public sector cybersecurity professionals moving into infrastructure-focused positions, college-level cybersecurity students seeking a practical understanding of incident response and infrastructure defense, and mid- to senior-level practitioners preparing to lead or support incident response in essential service environments. It is also written for security managers, incident responders, and technical leaders who need an operations-focused understanding of cyber defense across both enterprise IT and OT-dependent organizations.

Critical infrastructure cyber defense is often discussed in broad terms, but it is rarely explained in the way it is actually experienced. In theory, incident response is presented as a sequence of phases, supported by plans, policies, frameworks, and technical controls. In practice, incidents unfold amid competing priorities, incomplete information, staffing limitations, operational risk, public consequences, and leadership pressure. That gap between formal structure and real execution becomes especially important when the systems at risk support water, energy, transportation, fuel, emergency services, or other essential functions. In those environments, a poor cyber decision can quickly become an operational problem, and a technically correct decision can still be the wrong one if it creates unacceptable consequences for continuity, safety, or public trust.

The purpose of this book is not to replace technical manuals, engineering references, legal guidance, or incident-specific operational procedures. It is also not a certification guide or a checklist-based handbook. Its purpose is to help readers think clearly about how incident response works in real critical infrastructure environments, where speed and accuracy compete, where “secure” does not always mean “operationally viable,” and where defenders must often make decisions before certainty arrives. The focus throughout this book is on applied judgment, consequence-aware action, and the realities of defending systems that matter beyond the organization that owns them.

This book also reflects a transition in professional identity. Many readers entering critical infrastructure defense bring strong experience from military, government, public sector, or enterprise cybersecurity roles. Those experiences matter. Discipline matters. Calm under pressure matters. Structured thinking matters. At the same time, critical infrastructure defense introduces a broader and more complex operating environment. Authority is often more distributed. Ownership is often more fragmented. Legal, regulatory, contractual, operational, and public-facing concerns can all shape the same decision. The defender must learn not only how to manage technical risk, but how to operate effectively inside systems of consequence, dependency, and constraint.

Another reason this book exists is that too much cybersecurity writing remains disconnected from operational reality. Some writing leans so heavily into technical detail that it loses the decision environment. Some writing remains so abstract that it becomes difficult to apply. Some treat critical infrastructure as if it were simply enterprise IT with higher stakes. That approach is inadequate. Critical infrastructure defense requires a wider lens. It demands an understanding of mission continuity, process stability, public consequence, institutional friction, leadership behavior, and long-term resilience. It also requires humility. Defenders in these environments do not protect abstractions. They protect functions people depend on, often without ever seeing the systems or professionals behind them.

The chapters that follow are organized to move from the reality of incident response to the transition into infrastructure-focused roles, to the incident response lifecycle as it is actually experienced, and then into the specific challenges of critical infrastructure, leadership, and long-term resilience. The goal is to provide the reader with a professional, readable, and realistic guide that supports both learning and application. Some readers will approach this material as students preparing to enter the field. Others will bring years of operational experience. Both groups should find value here. The intention is to keep the writing accessible without becoming simplistic and to keep the material rigorous without losing practicality.

This is a serious field, and it deserves serious treatment. Critical infrastructure defense is not glamorous work. It is disciplined, often frustrating, and frequently invisible when done well. It requires people who can think clearly when the facts are incomplete, communicate honestly when pressure is high, and hold technical, operational, and institutional realities together without losing sight of what matters most. If this book helps readers do that more effectively, then it has served its purpose.

About This Book Series

Changing Lanes: A Transition into Critical Infrastructure Cyber Defense is part of a broader instructional series designed to support both professional study and applied learning. The main volume is intended to serve as the core text. Its purpose is to explain the principles, decision models, operational realities, and leadership considerations that shape incident response and cyber defense in critical infrastructure environments.

To support practical application, this book is accompanied by additional workbooks developed for structured, scenario-based learning. These companion resources are designed to help readers move from conceptual understanding to applied judgment. Rather than embedding a single narrative scenario throughout the core text, the broader series separates instruction from exercise so that readers can study the material cleanly in the main volume and then apply it in more focused, operationally realistic formats.

The companion materials include participant and facilitator workbook editions. The participant workbooks are intended for individual reflection, guided practice, team exercises, classroom use, and professional development activities. They present scenario-based problems, discussion prompts, decision points, and structured exercises that reinforce the concepts introduced in the core text. The facilitator workbooks are designed to support instructors, team leaders, training coordinators, and exercise organizers who need additional guidance for conducting discussions, tabletop exercises, and applied learning sessions built around the same material.

Because critical infrastructure defense occurs across a wide range of sectors, organizations, and operational conditions, the workbook series uses different scenarios rather than relying on a single fictionalized case throughout the main text. This approach allows readers and facilitators to examine multiple forms of pressure, including cyber incidents affecting utilities, operational technology environments, public-sector coordination challenges, continuity-of-operations concerns, leadership decision-making, and compound-crisis conditions. Together, the main text and workbook editions are intended to provide a more complete learning structure: one focused on understanding, the other on application.

Table of Contents

Part I. The Reality of Incident Response

Incident response is often described as a structured sequence, but in practice, it is rarely experienced that way. Real incidents begin in ambiguity, unfold under pressure, and force organizations to make decisions before the facts are fully settled. This section establishes the operational reality of incident response by examining the gap between formal models and lived execution, the uncertainty that shapes early decisions, and the threat conditions that make modern incidents difficult to identify and manage. The purpose of this part is to ground the reader in the real environment of cyber defense before moving into role transition, lifecycle execution, and critical infrastructure-specific challenges.

Chapter 1: Incident Response Is Controlled Chaos

Incident response is often introduced as a sequence of structured phases such as detection, analysis, containment, eradication, recovery, and post-incident review.[1] That structure is useful because it creates a common vocabulary for planning, coordination, and accountability. It helps organizations define responsibilities, develop procedures, and explain the broad progression of response activity. Without some kind of shared structure, incident response tends to become inconsistent, overly improvised, or dependent on the habits of a few individuals rather than on organizational discipline.

The problem is not that this structure exists. The problem is that many people come to mistake the structure for the lived reality of how incidents actually begin. In practice, incidents rarely present themselves in a clean, orderly way. They do not wait for the organization to gain clarity before generating consequences. They do not arrive at a convenient time. They do not respect staffing assumptions, support availability, or ideal escalation pathways. Instead, they usually emerge as fragments. A system behaves strangely but not catastrophically. An account does something unusual but not obviously malicious. An alert appears, but at a confidence level too low to command immediate agreement. An operator or administrator notices something that feels wrong but cannot yet explain why.

That is why the earliest phase of incident response is so often misunderstood. The beginning of a real incident is rarely defined by certainty. It is defined by ambiguity. An organization does not enter response mode because every fact is already known. It enters response

---

[1] Alexander Nelson, Sanjay Rekhi, Murugiah Souppaya, and Karen Scarfone, *Incident Response Recommendations and Considerations for Cybersecurity Risk Management: A CSF 2.0 Community Profile*, NIST SP 800-61 Rev. 3 (Gaithersburg, MD: National Institute of Standards and Technology, April 2025), https://doi.org/10.6028/NIST.SP.800-61r3.

mode because the available facts have become concerning enough that responsible people can no longer treat the situation as routine. The responder's first task is therefore not to know everything immediately. The first task is to recognize when uncertainty itself has become operationally significant.

This point matters because many practitioners, especially those who are newer to the field, assume that strong incident response begins with rapid technical confidence. They expect the responder to identify quickly whether an event is malicious, accidental, severe, or harmless. Real environments rarely provide that luxury. A strong responder is not the person who reaches certainty first. A strong responder is the person who can function responsibly before certainty arrives. That means recognizing when a pattern of weak indicators deserves structure, preservation, and disciplined escalation even though the full explanation is still incomplete.

That distinction is more difficult than it sounds. Most organizations live with a constant background level of irregularity. Systems slow down. Alerts produce noise. Administrators connect at unusual times. Maintenance happens imperfectly. Users make mistakes. Processes behave inconsistently for reasons that turn out to be benign. Because this is normal, defenders are trained, often for good reason, not to overreact to every anomaly. Serious incidents exploit that reality. They often begin in ways that remain individually explainable long enough to invite delay. A monitoring console may produce a weak signal. A process irregularity may be attributed to instability. An unusual access pattern may be rationalized as support activity. Each signal, by itself, remains manageable. The risk emerges when no one notices that the fragments are beginning to form a meaningful pattern.

This is why incident response should not be treated as the management of technical events alone. In its earliest stage, it is the management of uncertainty, consequences, and disciplined judgment. The organization must decide whether the available indicators justify additional scrutiny, whether they must be preserved, who should be involved, what assumptions should be avoided, and what level of structure is needed to prevent confusion from becoming the dominant condition. Those are not side issues. They shape whether the technical response that follows will begin on strong ground or weak ground.

A common mistake in organizations is to believe that action under uncertainty is somehow irresponsible. In reality, early response actions often do not require complete confidence. An organization can validate whether maintenance was scheduled, confirm the owners of systems and log sources, preserve volatile information, and quietly elevate concern without claiming to know more than it does. These early actions are not overreaction. They are the disciplined steps that preserve options while the technical picture is still developing. CISA's incident response guidance reflects this broader view by treating incident response planning as something that supports an organization before, during, and after a security incident rather than only after a threat has been fully confirmed.[2]

This is one reason incident response becomes a leadership problem much earlier than many professionals expect. Leadership at this stage does not necessarily mean an executive crisis meeting. It may involve a technical lead who prevents fragmented warning signs from remaining fragmented. It may involve an operations manager who insists that unexplained process behavior be treated seriously enough to investigate. It may involve a responder who frames the moment correctly for the organization: not as a confirmed catastrophe, but as a potentially significant event that justifies coordinated attention. Before an organization

---

[2] Cybersecurity and Infrastructure Security Agency, *Incident Response Plan (IRP) Basics* (Washington, DC: U.S. Department of Homeland Security, January 31, 2024), https://www.cisa.gov/resources-tools/resources/incident-response-plan-irp-basics.

can contain, eradicate, or recover, it must first recognize that the issue deserves structure. That recognition is often the earliest meaningful act of leadership in the entire incident.

This becomes even more important when the environment supports critical functions. In ordinary enterprise settings, an early misinterpretation may still be costly. It may lead to delayed containment, financial loss, customer harm, or reputational damage. Those consequences are serious. In critical infrastructure environments, however, the same delay may affect continuity of essential services, coordination across dependent systems, and the trust that public or operational stakeholders place in those services. This does not mean that every anomaly in a critical environment should be treated as a regional emergency. It does mean that responders must think beyond local symptoms when judging significance. A smaller system may still matter greatly if it supports a process, service, or dependency that extends beyond the local boundary.[3]

This broader view aligns with the way NIST and CISA frame cybersecurity as part of overall risk management rather than as a purely technical defensive function.[4] Incident response is not an isolated specialty operating outside the mission context. It exists within the broader context of how an organization understands, prioritizes, and manages risk. That means the responder must account not only for what is technically unusual, but also for what is operationally important. A weak indicator attached to a low-consequence system may deserve one level of scrutiny. The same indicator in a system tied to water, fuel, emergency support, communications, industrial control, or public continuity may deserve

[3] Kevin Stouffer et al., *Guide to Operational Technology (OT) Security*, NIST SP 800-82 Rev. 3 (Gaithersburg, MD: National Institute of Standards and Technology, September 2023).

[4] National Institute of Standards and Technology, *The NIST Cybersecurity Framework (CSF) 2.0*, NIST CSWP 29 (Gaithersburg, MD: National Institute of Standards and Technology, February 26, 2024).

another. The difference is not the signal alone. The difference is the environmental consequence in which the signal appears.

This is also why timing matters. Incidents do not occur under ideal staffing conditions simply because organizations would prefer that they do. They occur on nights, weekends, holidays, during storms, during turnover, and during exactly the kinds of periods when support depth is reduced. Reduced staffing does not create incidents, but it changes how incidents are perceived, how quickly they are validated, and how much ambiguity they are allowed to retain before someone acts. A weak signal that might have been clarified quickly under full staffing may remain unresolved much longer when fewer people are available, support relationships are slower, and ownership is harder to confirm. Timing is not background scenery. It is part of the incident condition.

That is especially true in operational technology and critical infrastructure environments. NIST's guidance on OT security makes clear that these environments are shaped by unique performance, reliability, and safety requirements that differ from many ordinary IT environments.[5] In practical terms, that means the responder is often working in systems where stability matters, process continuity matters, and the consequences of interruption are not confined to data loss or office productivity. Even when Chapter 1 is not yet focused on OT response in depth, it is important to establish early that incident significance in these environments cannot be measured only by what appears on a screen. Operational consequences may be broader than the first visible symptom suggests.

At the beginning of an incident, the organization's first objective is therefore not certainty. It is control. Control does not mean forcing a conclusion that the evidence cannot support. It does not mean dramatizing

---

[5] Cybersecurity and Infrastructure Security Agency, "Planning: Response & Recovery," accessed May 20, 2026, https://www.cisa.gov/planning-response-recovery

a situation before it is understood. It means reducing confusion enough that the organization can think clearly, preserve options, and avoid two equally damaging errors: ignoring a meaningful signal because it is inconvenient, or overreacting to an ambiguous signal without disciplined framing. Control may involve clarifying ownership, identifying what information matters most to validate first, protecting evidence, and ensuring that the issue is framed in language leadership can act on. These are foundational tasks, not secondary ones.

This framing also helps explain why incident response often feels frustrating even for experienced professionals. Early response work is rarely satisfying in a dramatic sense. It does not usually begin with visible proof, simple attribution, or perfect technical confidence. It begins with contradiction. Logs may not align. Different teams may each hold only part of the picture. Operators may report irregular behavior without knowing whether the cause is cyber, procedural, or mechanical. Administrators may observe access that is technically valid but contextually wrong. Leaders may ask for confidence levels that the technical team cannot honestly provide. The organization has not yet reached the point where anyone can say exactly what happened. It has only reached the point where the absence of a good explanation has become significant enough to demand disciplined attention.

That is the real beginning of incident response. It is earlier, less orderly, and more dependent on judgment than many formal diagrams suggest. A responder who understands this will be better prepared for the rest of the lifecycle because they will understand that response does not begin when the incident becomes obvious. It begins when the organization must decide whether weak but unexplained indicators deserve structure before the cost of waiting becomes greater than the cost of disciplined action.

The first phase of incident response is not about knowing everything. It is about recognizing enough, early enough, to avoid losing control.

Critical Thinking Question: When an organization sees several weak but unexplained indicators, what separates disciplined early action from overreaction, and how should a responder decide when ambiguity has become operationally significant?

Chapter 2: The Gap Between Theory and Execution

Most organizations that take cybersecurity seriously can produce some form of incident response plan.[6] They have policies, role descriptions, escalation paths, contact lists, severity definitions, and diagrams that describe how an incident should move through the organization. These materials matter because they provide structure, accountability, and a shared understanding of what response is supposed to look like. An organization with no defined response process is usually at a real disadvantage before an incident even begins.

The problem is that the existence of a plan is often mistaken for evidence of readiness. That mistake becomes visible only when a real incident begins to stress the assumptions built into the process. A document can describe a response model clearly and still fail to produce coordinated action under pressure. A plan can be technically complete and still be operationally weak. This is one of the most common and least appreciated problems in incident response. Organizations frequently measure whether the process exists, but they do not always measure whether the process can survive real conditions.

This distinction matters because real incidents do not unfold inside controlled planning assumptions. They do not occur only during full staffing, normal business hours, and ideal coordination conditions. They begin when people are unavailable, when information is incomplete, when vendors are slow to respond, when key owners are not immediately reachable, and when teams are forced to interpret events before they are comfortable doing so. Under those conditions, the difference between

---

[6] Alexander Nelson, Sanjay Rekhi, Murugiah Souppaya, and Karen Scarfone, *Incident Response Recommendations and Considerations for Cybersecurity Risk Management: A CSF 2.0 Community Profile*, NIST SP 800-61 Rev. 3 (Gaithersburg, MD: National Institute of Standards and Technology, April 2025), https://doi.org/10.6028/NIST.SP.800-61r3.

documented readiness and operational readiness becomes obvious very quickly.

Documented readiness is relatively easy to build. It can be assigned, reviewed, updated, and measured against formal requirements. A plan can be written to include the right sections, the right approval chain, and the right lifecycle terminology. Contact lists can be appended. Severity levels can be defined. Governance expectations can be met. These are useful activities, but they do not guarantee performance. Operational readiness is harder. It asks a different question. It asks whether the organization can still function coherently when the plan is no longer being tested by theory but by friction.

Friction is what turns a good-looking response process into a real capability test. The primary contact listed in the plan may not answer. The secondary contact may be outdated or unreachable. A vendor relationship that appears strong in policy language may collapse into a generic support queue during a weekend or holiday. A log source that everyone assumes is available may not be accessible to the people who need it most. A threshold for incident declaration may appear clear on paper but become difficult to interpret once the evidence is partial and the operational consequences of escalation are no longer theoretical. None of these failures is dramatic by itself. Together, they create the kind of delay and confusion that allow incidents to grow while the organization is still trying to align itself.

This is why plans often flatter organizations more than they protect them. They tend to reflect how the organization believes it will function when roles are current, dependencies are known, staffing is stable, and the environment behaves roughly as expected. Real incidents expose whether those assumptions still hold. If they do not, the incident response process can quickly become part of the problem rather than part of the solution.

Frameworks and lifecycle models should not be dismissed because of this. They remain necessary.[7] They create shared language, improve consistency, and support coordination across technical, operational, and leadership functions. The mistake is not in using them. The mistake is in assuming that the framework itself produces execution quality. A response framework is a map. It is not the same thing as the organization's ability to move under pressure. A team can know the phases of incident response perfectly and still perform poorly if ownership is unclear, support relationships are brittle, communications are slow, or decision thresholds become unstable as soon as ambiguity rises.

This gap between theory and execution is particularly important in critical infrastructure environments. In these settings, the consequences of delay are not limited to extended technical exposure. Response weakness can affect continuity of service, coordination across dependent systems, public confidence, and the organization's ability to make disciplined decisions while still preserving operations.[8] A plan that looks acceptable in a general enterprise context may prove inadequate when the systems involved support water, energy, transportation, communications, public safety, or other essential functions. In those environments, the margin for procedural confusion is smaller because the consequences of losing control can spread beyond the local team or facility.

One reason this problem persists is that organizations often use the wrong measures of preparedness. It is easier to confirm that a plan exists

---

[7] National Institute of Standards and Technology, *The NIST Cybersecurity Framework (CSF) 2.0*, NIST CSWP 29 (Gaithersburg, MD: National Institute of Standards and Technology, February 26, 2024).

[8] Kevin Stouffer et al., *Guide to Operational Technology (OT) Security*, NIST SP 800-82 Rev. 3 (Gaithersburg, MD: National Institute of Standards and Technology, September 2023).

than to confirm that it will work well. It is easier to document that a tabletop exercise occurred than to determine whether that exercise genuinely exposed weak assumptions. It is easier to verify that a role has been assigned than to determine whether the person in that role can be reached quickly, has the authority expected, and understands how to perform under real pressure. Administrative completion is not meaningless, but it should never be confused with tested capability.

A well-run incident response program, therefore, requires more than a plan. It requires maintained contacts, validated access paths, clear ownership, realistic escalation criteria, accessible logging, and response relationships that still function when staffing is thin or conditions are degraded.[9] It also requires that people across the organization understand not only what the plan says, but how the plan will actually be used when the available evidence is ambiguous, and the organization must decide whether to act before certainty arrives. In other words, it requires the organization to prepare for friction, not merely to document intent.

This is one reason realistic exercises matter so much. An exercise that simply walks participants through a neat lifecycle sequence without introducing uncertainty, communication difficulty, unavailable personnel, vendor delay, or decision friction may create confidence without improving readiness. A useful exercise is not one that proves the organization can recite the process. It is one that reveals where the process begins to fail under pressure. That kind of exercise is less flattering, but far more valuable. It forces the organization to confront the real question: if this happened under inconvenient conditions, would the process still hold?

The same principle applies to leadership expectations. Many leaders assume that if the organization has approved plans and assigned

---

[9] Cybersecurity and Infrastructure Security Agency, *Incident Response Plan (IRP) Basics* (Washington, DC: U.S. Department of Homeland Security, January 31, 2024), https://www.cisa.gov/resources-tools/resources/incident-response-plan-irp-basics.

responsibilities, the process will naturally activate when needed. In practice, activation often depends on interpretation. Someone must decide whether the event is serious enough to escalate. Someone must determine whether the available indicators justify involving legal, operations, executives, or external partners. Someone must decide whether the issue is still a technical irregularity or now a coordinated organizational problem. Those decisions are rarely made with complete information. If the plan assumes more clarity than reality provides, the organization may lose valuable time while debating what phase it is in rather than improving its understanding of the actual risk.

This is where the difference between formal process and operational capability becomes most visible. A formal process tells the organization what it is supposed to do. Operational capability allows it to do that work while conditions are degraded. The former is necessary. The latter is decisive. A mature organization understands that both matter and does not allow governance completion to substitute for tested performance.

For students and newer practitioners, this is an important lesson because early professional education often emphasizes frameworks, standards, and lifecycle diagrams. That emphasis is useful as a foundation, but it can create the false impression that knowing the structure is equivalent to being ready to execute inside it. It is not. Response quality depends on whether the structure holds when communications are slower than expected, facts are incomplete, and decisions carry real consequences. That is why incident response cannot be understood only as a formal sequence. It must also be understood as a test of organizational function.

For experienced practitioners, the lesson is usually learned more bluntly. The most frustrating response problems are often not caused by the attacker alone. They are caused by stale assumptions inside the organization's own process. A critical owner is not reachable. A support path is weaker than expected. A vendor cannot move at the tempo required. Logging exists, but not in a way the team can use immediately. Severity definitions are written, but not interpreted consistently enough to drive quick action. These are not failures of awareness. They are failures of readiness.

The phrase "gap between theory and execution" should therefore not be read as a criticism of planning. Planning is necessary and should be taken seriously. The point is more demanding than that. A plan must be treated as the beginning of readiness, not the proof of it. If the organization has not tested the process against ambiguity, degraded staffing, vendor friction, and imperfect communication, then it has not yet shown that the process is operationally sound.

In the end, incident response does not fail only because attackers are skilled. It often fails because organizations discover too late that the weaknesses were already present inside their own response mechanisms. A documented process may satisfy governance expectations, but only an operationally durable process can support the organization when the incident begins to apply pressure. The difference between those two states is where theory ends, and execution begins.

Critical Thinking Question: If an organization has a documented incident response plan but struggles to execute it under real-world pressure, where is the true weakness most likely to be found: in the plan itself, in the assumptions behind the plan, or in the organization's understanding of what readiness actually means?

Chapter 3: The Modern Threat Landscape Facing Responders

Many organizations still carry an outdated mental model of what a serious cyber incident looks like. In that model, attacks are noisy, perimeter-focused, visibly malicious, and easy to separate from ordinary administrative activity. Malware is expected to look foreign, access attempts are expected to be obviously suspicious, and the beginning of an incident is expected to feel different enough from normal operations that the organization will recognize it quickly. That model is no longer reliable.

Modern incidents are often difficult not because they are always technically spectacular, but because they move through places the organization already trusts.[10] Administrative tools, remote access pathways, valid credentials, vendor relationships, shared services, cloud platforms, and legitimate system functions can all become part of the attack path. The adversary does not always need to break every door down. In many cases, the adversary succeeds by using the doors that are already open, or by making their activity resemble authorized behavior closely enough that defenders hesitate to classify it as malicious.

That shift matters because it changes what responders should pay attention to in the earliest stages of an incident. If defenders are still looking primarily for obvious malware, obvious scanning, or clearly hostile intrusion patterns, they may miss the more dangerous indicators. A privileged account used at an unusual time may matter more than a blocked malware sample. A remote connection through a known support channel may deserve more scrutiny than a noisy failed login pattern. A sequence of technically valid actions can become a stronger indicator of compromise than a single obviously suspicious event. The threat

---

[10] National Institute of Standards and Technology, *Zero Trust Architecture*, NIST SP 800-207 (Gaithersburg, MD: National Institute of Standards and Technology, August 2020).

landscape has not become simpler. It has become more dependent on identity, trust, and the abuse of normal pathways.

This is one reason modern incidents can remain quiet for so long. The attacker is not always trying to create immediate disruption. In many cases, the more useful strategy is to establish access, learn the environment, identify dependencies, and preserve the ability to move later without attracting early response. That does not require cinematic sophistication. It requires patience, discipline, and a good understanding of how the target environment normally operates. When an adversary can move through valid accounts, legitimate remote services, or approved administrative mechanisms, the resulting activity may not look extraordinary enough to trigger rapid consensus. It may only look slightly wrong, and that slight wrongness is where mature response begins.

Ransomware provides a useful example of how outdated thinking can distort response. Many people still treat ransomware as though the visible encryption event is the whole incident. In reality, that visible disruption is often the final stage of a much longer compromise.[11] By the time systems are encrypted, data is stolen, or extortion demands appear, the attacker may already have established persistence, mapped the environment, identified key dependencies, and gained access through multiple paths. Focusing only on the visible end-stage can cause responders to underestimate the larger access problem that made the ransomware event possible in the first place.

This is why responders should be careful not to define an incident too narrowly based on its most visible symptom. The organization may discover a ransomware note, a service outage, or a set of disrupted systems and still misunderstand the actual problem. The deeper issue

---

[11] Cybersecurity and Infrastructure Security Agency et al., *#StopRansomware Guide* (Washington, DC: U.S. Department of Homeland Security, updated March 2025), https://www.cisa.gov/stopransomware/ransomware-guide.

may be compromised identities, abused trust relationships, weak remote access controls, poor network access assumptions, or a third-party dependency that allowed the adversary to move quietly before disruption began. If responders focus only on the final expression of the incident, they risk containing the symptom while leaving the enabling conditions largely intact.

Identity is now central to this problem.[12] Modern attackers do not need to rely only on malware persistence or network exploitation when valid accounts, service credentials, privileged identities, and federated trust relationships can provide lasting access. A compromised administrator account, a poorly governed service account, or a trusted remote support identity can create opportunity that survives even after visible malicious tooling has been removed. In practical terms, this means the threat landscape is no longer defined only by software artifacts or network events. It is increasingly defined by who or what the environment is willing to trust, and under what conditions that trust can be abused.

This identity-centered reality has major implications for responders. It means the question "What malware is present?" is often less important at the outset than questions such as "What identities have been exposed?" "What privileged actions were taken?" "What access was technically valid but contextually abnormal?" and "What trust relationships allowed that access to matter?" Those questions are harder, because they require defenders to understand not only technical telemetry, but normal administrative behavior, account purpose, support relationships, and operational context. They also require more discipline because technically valid activity is often easier to rationalize than clearly unauthorized behavior.

---

[12] National Institute of Standards and Technology, *Zero Trust Architecture*, NIST SP 800-207 (Gaithersburg, MD: National Institute of Standards and Technology, August 2020).

Third-party access and inherited trust create additional complexity. Vendor support channels, managed services, cloud administration, identity federation, and remote maintenance relationships are often necessary parts of modern operations. They are not optional conveniences that organizations can simply wish away. At the same time, these relationships create attack opportunities that many organizations fail to evaluate honestly enough. CISA's guidance on modern network access security and on commonly exploited initial access weaknesses reflects this reality by emphasizing that weak controls around remote access, identity, and externally reachable services are routinely exploited by adversaries.[13] The danger here is not the existence of support relationships themselves. The danger is the assumption that because those relationships are legitimate, they are therefore low risk.

For responders, this means that "trusted" should never be treated as synonymous with "safe." A trusted path may still be the path an attacker uses most effectively. A managed service connection may still become part of a compromise. A support account may still be the wrong place to place unexamined confidence. The threat landscape rewards attackers who can exploit this gap between legitimacy and scrutiny.

Critical infrastructure environments raise the stakes even further. Many of these environments contain legacy systems, mixed IT and OT dependencies, support relationships that are difficult to change quickly, and continuity requirements that make aggressive disruption costly. This does not mean critical infrastructure defenders are inherently behind or incapable. It means the environments they protect often include more

---

[13] Cybersecurity and Infrastructure Security Agency et al., "Weak Security Controls and Practices Routinely Exploited for Initial Access," May 17, 2022, https://www.cisa.gov/news-events/alerts/2022/05/17/weak-security-controls-and-practices-routinely-exploited-initial-access; Cybersecurity and Infrastructure Security Agency et al., *Modern Approaches to Network Access Security*, June 18, 2024, https://www.cisa.gov/resources-tools/resources/modern-approaches-network-access-security.

complex combinations of operational necessity, institutional friction, and trust dependency than ordinary enterprise settings. When an attacker enters such an environment through valid accounts, vendor pathways, or weakly governed remote access, the organization may face greater difficulty in recognizing the threat early and greater consequences if recognition comes late.

This broader consequence environment is why responders need to think in terms of dependencies rather than isolated systems. A smaller node in a larger infrastructure environment may not matter because of its size. It may matter because of what depends on it, what it can reach, or what trust relationships connect it to broader services. An identity event in one segment may have significance far beyond that segment if it touches administrative access, remote support, or operational continuity. A defender who focuses only on the compromised host or the visible alert may miss the more important question: what else does this trust relationship make possible?

The modern threat landscape, therefore, demands a different kind of professional attention. Responders must become better at asking whether activity is merely valid in technical terms or also reasonable in operational terms. They must be slower to dismiss behavior just because it passed through an approved channel. They must recognize that quiet incidents are not necessarily smaller incidents. They must understand that the absence of obvious disruption often means the adversary is still operating in the portion of the lifecycle where patience is more useful than noise.

This is not a reason for pessimism. It is a reason for precision. Defenders do not need to become dramatic or speculative in response to this environment. They need to become more disciplined. They need to pay closer attention to identity, access, trust, timing, dependency, and context. They need to understand that the technical legitimacy of an action does not answer the question of whether the action makes sense in the environment where it occurred. Most of all, they need to stop assuming that serious incidents will identify themselves through obvious hostility before they become expensive.

A modern responder should therefore think less in terms of "What attack are we looking at?" and more in terms of "What access is being exercised, what trust is being abused, and what opportunity does that create if we are wrong about what we are seeing?" That is the more useful entry point into the threat landscape that defines real incident response today.

Critical Thinking Question: Why are legitimate-looking access patterns, administrative actions, and trusted support pathways often more dangerous in the early phase of an incident than obviously malicious activity, and what does that imply for how responders should interpret weak signals?

Part II. Transitioning Into Critical Infrastructure Defense

Moving into critical infrastructure defense is not a matter of abandoning prior experience. It is a matter of adapting strong cybersecurity habits to a more constrained, more interdependent, and more consequence-driven environment. This section examines the transition from cyber warfighter to cyber defender, the challenge of decision-making under uncertainty, the realities of shared authority and civilian leadership, and the legal and operational boundaries that shape action in essential-service environments. The purpose of this part is to help the reader understand that technical competence remains necessary, but context, restraint, and consequence-aware judgment become even more important in this field.

Chapter 4: Changing Lanes Without Losing Your Edge

Transitioning into critical infrastructure cyber defense is not the same thing as starting over. Many readers entering this field from military cyber operations, public-sector service, enterprise security, or related technical roles already possess habits that translate well. Discipline under pressure translates well. Respect for the mission translates well. Escalation awareness, structured thinking, operational planning, and the ability to function in imperfect conditions all translate well.[14] These are not minor advantages. In many infrastructure environments, they are exactly the qualities that keep a response effort from deteriorating into noise and drift.

The transition becomes difficult not because prior experience loses value, but because the operating environment changes. Critical infrastructure defense is shaped by a broader and more fragmented decision space than many readers initially expect. In some professional settings, especially those with clearer authority structures, it is possible to assume that technical urgency will move directly through a chain of command and produce action. In infrastructure environments, that assumption is often wrong. A technically sound recommendation may still have to move through operations leadership, engineering staff, legal counsel, executives, outside vendors, compliance personnel, regulators, insurers, public communications officials, or external partners before it becomes an executed decision. The issue is not that these environments are unserious. The issue is that they are more distributed, more interdependent, and more constrained by consequence.

That difference can frustrate experienced professionals who are used to more centralized decision structures. A defender may identify a

---

[14] Roger C. Petersen et al., *Workforce Framework for Cybersecurity (NICE Framework)*, NIST SP 800-181 Rev. 1 (Gaithersburg, MD: National Institute of Standards and Technology, November 2020).

real technical risk and still be unable to act on it unilaterally because the affected system supports a physical process, a contractual support pathway, or a public service function. A responder may understand the cyber consequence immediately while other stakeholders evaluate the same situation in terms of continuity, legal exposure, operational impact, or public confidence. In these settings, authority is not absent. It is shared. That changes how technical professionals must think if they want to remain effective.

This is one reason the phrase "changing lanes" is more accurate than "starting fresh." The reader is not abandoning prior strengths. The reader is moving those strengths into a new operating lane where the rules of motion are different. Structured thinking remains valuable, but it must now coexist with distributed ownership. Calm under pressure remains valuable, but it must now support coordination across people who do not all view the incident through the same lens. Mission focus remains valuable, but the mission is now framed not only by technical success, but also by service continuity, process stability, legal boundaries, and public consequence.

Critical infrastructure environments make this shift unavoidable because the systems being defended do not exist only for organizational convenience. They support essential services and functions that may affect communities, supply chains, emergency operations, or industrial continuity.[15] That fact changes how cyber decisions are evaluated. In an enterprise setting, a technical team may be able to recommend aggressive disruption if the security benefit is strong enough. In infrastructure environments, that same recommendation may require a more careful balancing of consequences. The defender must therefore learn not only

[15] Cybersecurity and Infrastructure Security Agency, "Critical Infrastructure Security and Resilience," accessed May 20, 2026, https://www.cisa.gov/topics/critical-infrastructure-security-and-resilience.

how to identify cyber risk, but how to explain that risk in terms the larger organization can act on.

This is where many transitioning professionals encounter their first real friction. They are often correct technically but insufficiently effective organizationally. They know what should be done from a security standpoint, but they have not yet adapted to the reality that civilian infrastructure environments often require influence, framing, and consequence-aware communication rather than direct authority. A responder who is used to sharper command lines may initially see this as weakness, bureaucracy, or indecision. Sometimes it is. More often, it is simply the normal consequence of operating in a setting where ownership, mission, and accountability are distributed across multiple legitimate stakeholders.

The concept of unified effort is useful here, even when formal unified command is not present.[16] In infrastructure defense, cyber teams, operations leaders, engineers, vendors, executives, and public-sector counterparts often have overlapping but nonidentical responsibilities. The challenge is not to erase those distinctions. The challenge is to work effectively across them. A responder who expects every stakeholder to evaluate the incident primarily through a cyber lens will misread the environment. A responder who can frame the issue in terms of service impact, operational tolerance, legal constraints, and mission priorities will usually gain more traction even when formal authority is limited.

This is one reason professional credibility becomes so important in civilian infrastructure roles. Technical skill matters, but technical skill by itself does not always move the organization. Credibility is built when a defender consistently demonstrates that they understand not only the

---

[16] Federal Emergency Management Agency, *National Incident Management System* (3rd ed., Washington, DC: U.S. Department of Homeland Security, July 2017).

cyber risk, but also the business, operational, and public consequences attached to the decision. Infrastructure leaders are more likely to act decisively when they believe the cyber professional understands the wider environment and is not simply pushing the most aggressive technical option out of habit. Trust grows when the responder shows discipline, accuracy, and an ability to translate technical conditions into decision-relevant language.

The NICE Workforce Framework is useful in this context because it emphasizes cybersecurity work as a set of roles, tasks, knowledge, and skills that can be adapted across sectors and environments rather than confined to one organizational model.[17] That matters for transitioning professionals because it reinforces an important point: cybersecurity competence is transferable, but it is not context-free. The same person may be highly capable in one environment and less effective in another if they do not adapt their communication, assumptions, and decision framing to the mission and structure around them. The transition into critical infrastructure defense is therefore not only a technical adjustment. It is a contextual adjustment.

That adjustment also affects how mission should be understood. Many readers entering this field already have a strong sense of mission, and that is an asset. However, critical infrastructure mission is not always framed in familiar command language. It may be expressed through continuity of water, continuity of fuel, continuity of transportation, continuity of emergency support, or the stable operation of systems the public rarely sees until they fail. The defender is still serving something larger than the network, but the language of that service may be more civilian, more operational, and more consequence-driven than some readers are used

---

[17] Roger C. Petersen et al., *Workforce Framework for Cybersecurity (NICE Framework)*, NIST SP 800-181 Rev. 1 (Gaithersburg, MD: National Institute of Standards and Technology, November 2020).

to. That does not make it less real. In many cases, it makes the stakes more visible because the connection between system failure and human consequence is harder to ignore.

This transition is also important for students and early-career practitioners. It is easy to imagine cybersecurity as a field where technical expertise alone determines effectiveness. That is not true in critical infrastructure defense. Technical skill is necessary, but it is not sufficient. The defender must understand how institutions function, how authority is distributed, how critical services depend on one another, and how to communicate with people who do not share the same technical vocabulary. A student preparing to enter this field should therefore think beyond tools and certifications. The more relevant question is whether they are learning how to operate in environments where consequences, coordination, and judgment matter as much as technical depth.

For many readers, the largest adjustment is learning that influence often matters as much as control. In some cyber roles, especially highly centralized ones, the defender can move directly from technical findings to technical action. In infrastructure roles, the defender may need to persuade, brief, coordinate, or sequence action through people who own adjacent parts of the problem. That does not mean the defender becomes less decisive. It means decisiveness must be expressed in a more disciplined way. Instead of saying only, "This is a cyber problem, and it must be shut down," the responder may need to say, "This is the risk, this is what it can affect, this is the cost of delay, this is the cost of disruption, and this is the least harmful path to action." That is not weaker leadership. It is more mature leadership.

NIMS guidance is helpful in understanding this broader coordination reality because it emphasizes that multiple organizations, sectors, and stakeholders often need to work together under domestic incident conditions.[18] Although critical infrastructure cyber incidents do not always activate formal emergency management structures, the underlying lesson still applies. Shared consequence does not automatically produce shared control. Effective action in distributed environments requires more than technical correctness. It requires structure, communication, and a disciplined respect for how responsibilities are divided.

None of this means that the transitioning professional should become passive or overly deferential. The goal is not to abandon directness or operational seriousness. The goal is to apply those strengths intelligently in an environment where technical truth is only one input into the larger decision process. A defender who loses discipline in order to become "more collaborative" has made the wrong adjustment. A defender who keeps discipline but learns how to align it with distributed authority has made the right one.

That is the real meaning of changing lanes without losing your edge. The edge is still valuable. The challenge is learning how to use it in a system where mission is broader, ownership is shared, and consequence extends beyond the technical domain. Professionals who make that shift successfully do not become less sharp. They become more effective.

Critical Thinking Question: Which qualities from a highly structured cybersecurity environment remain most useful in critical infrastructure defense, and which ones must be adapted if the responder is going to be effective in a setting where authority, mission, and consequence are shared across multiple stakeholders?

---

[18] Federal Emergency Management Agency, *National Incident Management System* (3rd ed., Washington, DC: U.S. Department of Homeland Security, July 2017).

Chapter 5: Fog of War and Cyber Decision-Making

One of the most persistent misconceptions in cybersecurity is the belief that good decisions are made only after the facts are clear. In practice, incident response rarely offers that kind of clarity early enough to be comfortable. Responders are often required to make decisions while the technical picture is incomplete, the scope is still evolving, and the consequences of delay are beginning to accumulate. This is not a failure of process. It is a defining condition of real-world incident response.

The phrase "fog of war" is often used loosely, but in cyber defense, it has a practical meaning. It refers to the condition in which the organization must act without full visibility into what is happening, what systems are affected, what access has been gained, or what the adversary intends to do next. Logs may be incomplete. Indicators may conflict. Different teams may hold different parts of the picture. A responder may know enough to believe the situation is serious without knowing enough to describe the incident with confidence. That is where decision-making begins in earnest.

This creates an uncomfortable but unavoidable truth. The organization will often need to decide before it knows as much as it would prefer. In some cases, waiting for certainty can be more damaging than acting on disciplined judgment. NIST's guidance on risk assessment is useful here because it frames organizational decision-making in terms of risk, impact, likelihood, and available courses of action rather than in terms of perfect knowledge.[19] In other words, the standard is not whether uncertainty has been eliminated. The standard is whether the organization

---

[19] Joint Task Force Transformation Initiative, *Guide for Conducting Risk Assessments*, NIST SP 800-30 Rev. 1 (Gaithersburg, MD: National Institute of Standards and Technology, September 2012), https://doi.org/10.6028/NIST.SP.800-30r1.

understands enough about the risk to justify a responsible course of action.

That is a critical distinction. Inexperienced responders often treat uncertainty as a reason to delay. Experienced responders are more likely to treat uncertainty as something to be managed explicitly. That does not mean they become reckless. It means they understand that ambiguity is part of the operating environment and must be addressed through disciplined reasoning rather than through wishful delay. A team that refuses to act until it is fully comfortable may discover that the incident has used that time more effectively than the defenders did.

This is why one of the most useful habits in incident response is separating what is known from what is suspected, inferred, or still unknown. The distinction sounds elementary, but many poor decisions begin when teams blur those categories under stress. A valid observation becomes an assumption without being labeled as one. An assumption becomes a working conclusion. A working conclusion becomes the basis for action before anyone has paused to ask how much of it is evidence and how much of it is interpretation. The result may be overconfidence, delay, or escalation in the wrong direction.

Good responders do not eliminate interpretation. Interpretation is unavoidable. They do, however, make interpretation visible. They ask which facts are confirmed, which patterns are emerging, which explanations remain plausible, and which uncertainties matter most to the next decision. That last point is especially important. Not every unknown carries equal weight. Some gaps in understanding can remain open while the organization proceeds. Others directly affect whether the next action is proportionate, safe, or strategically sound.

This is where decision quality and decision comfort begin to separate. A comfortable decision is one made when the team feels it finally has enough information to remove doubt. A good decision is one that is reasonable, proportionate, and aligned to the risk conditions at the time it must be made. Those two things are not always the same. In fact, in serious incidents, they often diverge. The more urgent the situation becomes, the less likely it is that the organization will enjoy the luxury of full confidence. Waiting for emotional comfort may simply mean waiting until the decision space narrows and the cost of the remaining options rises.

This does not mean speed is always preferable to caution. Acting too quickly can create its own damage. Overly aggressive containment can disrupt operations unnecessarily. Poorly framed escalation can generate confusion rather than clarity. A technical action taken on a weak assumption can destroy useful evidence or force the organization into a more expensive recovery path. The point is not that responders should move fast regardless of uncertainty. The point is that they must learn how to evaluate the cost of delays against the cost of error. That is a more demanding standard than either "act immediately" or "wait until sure."

The quality of that evaluation depends heavily on framing. A responder who presents leadership with only one recommendation and no explanation of trade-offs is not supporting good decision-making. A responder who presents nothing because the facts are incomplete is not supporting it either. The more useful approach is to explain what is known, what remains uncertain, what the principal risks are, what options exist, and what each option is likely to cost the organization. CISA's cyber incident response playbooks and related planning guidance reinforce this point by emphasizing structured coordination, defined roles, and operational procedures that allow organizations to move even when an incident is still being understood.[20]

This is where many technically strong teams struggle. They know what they would prefer to do from a cybersecurity standpoint, but they have not translated that preference into decision-grade language. Leaders generally do not need raw telemetry. They need to know whether the risk is growing, whether the issue appears isolated or expanding, what happens if the organization waits, what happens if it acts now, and which choices require broader coordination. A technically correct recommendation that is poorly framed may be less useful than a less detailed recommendation that clearly explains the consequences and urgency.

This is also why a common operating picture matters.[21] A response team does not need every stakeholder to hold the same level of technical detail, but it does need enough shared understanding that people are making decisions within the same incident rather than within competing interpretations of it. If operations believes the issue is still local, leadership believes it is only hypothetical, and technical teams believe it is already expanding, the organization may lose time to internal disagreement rather than to adversary action. Shared understanding is not the same thing as complete certainty. It is the disciplined alignment of the people who must make the next set of decisions.

---

[20] Cybersecurity and Infrastructure Security Agency, *Federal Government Cybersecurity Incident and Vulnerability Response Playbooks* (Washington, DC: U.S. Department of Homeland Security, updated 2024), https://www.cisa.gov/resources-tools/resources/federal-government-cybersecurity-incident-and-vulnerability-response-playbooks.

[21] Federal Emergency Management Agency, *National Incident Management System* (3rd ed., Washington, DC: U.S. Department of Homeland Security, 2017); see also Federal Emergency Management Agency, "National Incident Management System," updated July 28, 2025, https://www.fema.gov/emergency-managers/nims.

In practical terms, this means that good cyber decision-making often begins with a small set of disciplined questions. What do we know for certain? What do we think is happening, and why? What remains unknown that could materially change the next decision? What are the likely consequences of waiting? What are the likely consequences of acting now? Which decision belongs to the technical team, and which must be elevated because the consequences are broader than the technical domain alone? These questions do not remove uncertainty. They prevent uncertainty from controlling the organization by default.

This discipline is particularly important in critical infrastructure environments. The reason is not only that the stakes are higher. The reason is that the cost of both delay and error tends to be broader. A poor cyber decision may create operational disruption, continuity problems, public consequences, or downstream effects that are not visible to the initial responder. At the same time, waiting too long when the evidence is incomplete may allow the incident to spread to systems, processes, or dependencies that make later action much more costly. In these environments, defenders rarely choose between a clean option and a risky option. They often choose between two or more imperfect options, each carrying a different type of risk.

This is why disciplined decision-making should be evaluated by the quality of reasoning at the time, not solely by hindsight. After an incident, it is always easier to identify what would have been ideal once the full picture is known. That is useful for learning, but it can be misleading if it becomes the standard for judging every in-the-moment decision. A sound response decision is one that was reasonable given the known facts, the relevant uncertainties, the likely consequences, and the available options at that moment. This standard is not lenient. It is realistic. It holds responders accountable for disciplined judgment rather than for impossible foresight.

For readers transitioning into critical infrastructure defense, this chapter marks an important mental shift. The goal is not to become comfortable with uncertainty in a casual sense. The goal is to become competent inside it. That means learning how to reason, how to frame, how to communicate partial truth honestly, and how to avoid both premature certainty and avoidable delay. Responders who develop this skill do not remove fog from the environment. They prevent fog from becoming paralysis.

The practical lesson is straightforward, even if the work is not. A responder does not need perfect clarity to support a good decision. The responder does need enough discipline to distinguish fact from assumption, enough judgment to weigh the cost of waiting against the cost of acting, and enough communication skills to turn incomplete information into usable decision support. That is what real cyber decision-making looks like in the early and middle phases of an incident.

Critical Thinking Question: When the organization has enough evidence to believe a situation may be serious but not enough to feel certain, what standard should determine whether it acts now or waits for more clarity?

Chapter 6: Command Intent, Escalation, and Civilian Leadership

Once an incident moves beyond an isolated technical concern, the next challenge is no longer only technical analysis. The organization must decide who needs to know, who needs to decide, what must be protected first, and how the available information should be framed so that action becomes possible. This is the point where incident response becomes an organizational coordination problem as much as a cybersecurity problem. A technically correct observation is not enough by itself. It must be translated into a form that supports decision-making across leaders, operators, legal counsel, external partners, and others who may not share the same technical vocabulary or priorities.

In more centralized environments, professionals sometimes assume that technical urgency will quickly move through a chain of authority and result in action. In many civilian organizations, and especially in critical infrastructure environments, that assumption is unreliable. Authority is often distributed. Ownership is often shared. Operational, legal, regulatory, contractual, and public-facing consequences may all be attached to the same decision. That does not mean the organization lacks structure. It means the structure is broader, more interdependent, and more dependent on coordination than some responders initially expect.

This is why the idea of command intent remains useful even outside a formal command environment. The concept is simple. Before the organization can act effectively, it must understand which outcome matters most, which risks are acceptable, which constraints cannot be ignored, and which priorities should guide action when the picture remains incomplete. In other words, the organization must understand what it is trying to preserve and the kinds of harm it is willing to absorb to do so.

Command intent, in this sense, is not about importing military language for style. It is about clarifying purpose under pressure.

That clarification matters because incidents create a natural tendency toward fragmented urgency. The technical team may want immediate restriction of access or aggressive containment. Operations may want to preserve continuity for as long as possible. Legal may be concerned with regulatory obligations, contractual implications, or the wording used in internal and external communications. Executive leadership may want a defensible understanding of impact before authorizing disruptive action. External vendors may need to be consulted before changing supported pathways or systems. All of these positions may be legitimate. The problem is not that people disagree. The problem is that disagreement without structure can quickly become decision paralysis.

Effective escalation exists to prevent that outcome. Too many organizations treat escalation as if it were simply the act of informing more people. That is incomplete. Escalation is useful only when it enables decisions. A technically detailed message sent to leadership is not an effective escalation if the recipients still do not understand what is at stake, what must be decided, or what the consequences of delay may be.[22] In that sense, escalation is not a communications volume problem. It is a decision-support problem.

This distinction is critical because technical teams often believe they have escalated an issue when they have only transmitted information. Leadership, meanwhile, may believe it has been informed, yet still lacks the structure needed to act. The result is one of the most common and damaging failure patterns in incident response: everyone believes the

---

[22] Cybersecurity and Infrastructure Security Agency, *Incident Response Plan (IRP) Basics* (Washington, DC: U.S. Department of Homeland Security, January 31, 2024), https://www.cisa.gov/resources-tools/resources/incident-response-plan-irp-basics.

issue has been elevated, but no one has actually clarified the decision space. The technical team assumes the matter is now "with leadership." Leadership assumes the technical team is still "investigating." Meanwhile, the incident continues to develop.

A more disciplined model begins with a common operating picture. FEMA's National Incident Management System emphasizes the value of a common operating picture because decision-makers cannot coordinate effectively if they are acting inside different versions of the same event.[23] In cyber response, a common operating picture does not mean that everyone receives identical technical details. It means that the people who must act share enough understanding of the issue, its likely consequences, and its current status that they can make decisions from the same reality rather than from competing interpretations. Without that alignment, technical accuracy at the team level does not reliably translate into organizational coherence.

This is why responders must learn to frame an incident for multiple audiences without distorting the underlying truth. A technical team may need indicators, system details, access patterns, and uncertainty boundaries. An executive decision-maker may need to know what services or functions are at risk, whether the issue appears to be localized or expanding, what options are available, and what decisions are needed now versus later. Operations leadership may need to understand what actions could disrupt continuity. Legal and communications staff may need precise, disciplined language to ensure the organization neither overstates nor understates what is known. The responder's role in this environment is

---

[23] Federal Emergency Management Agency, *National Incident Management System* (3rd ed., Washington, DC: U.S. Department of Homeland Security, 2017); see also Federal Emergency Management Agency, "National Incident Management System," updated July 28, 2025, https://www.fema.gov/emergency-managers/nims.

not to falsely simplify the incident. It is to structure it honestly for the people who must help manage it.

This is where civilian leadership reality becomes more complex than many responders initially expect. In critical infrastructure environments, technical truth rarely moves alone. It often arises alongside concerns about continuity, public consequences, regulatory exposure, vendor obligations, and external dependencies. CISA's incident response planning guidance makes this clear by emphasizing that a workable incident response capability must clarify roles, responsibilities, communications, and coordination before the organization is under full pressure.[24] The reason is straightforward. By the time the issue becomes serious enough to demand broad attention, the organization can no longer afford to discover basic coordination assumptions in real time.

Decision rights are especially important in this context. Every incident contains decisions that belong at different levels. Some can and should remain inside the technical team. Others require an operational agreement because the proposed action may affect process continuity, service delivery, or safety. Still others require executive involvement because the impact reaches beyond technical systems into organizational risk tolerance, external reporting, or public consequence. Confusion about who owns which decision is one of the fastest ways to lose time. A responder who does not understand decision rights may escalate too late, escalate too broadly, or recommend technical action in a space where the authority to act is not theirs to exercise alone.

This is one reason why command intent and escalation must be tied together. Command intent clarifies what matters. Escalation moves the

---

[24] Cybersecurity and Infrastructure Security Agency, *Federal Government Cybersecurity Incident and Vulnerability Response Playbooks* (Washington, DC: U.S. Department of Homeland Security, updated 2024), https://www.cisa.gov/resources-tools/resources/federal-government-cybersecurity-incident-and-vulnerability-response-playbooks.

issue into the right decision space. Together, they prevent the response from becoming a contest between technical urgency and organizational hesitation. Without them, teams tend to default to one of two unhelpful patterns. In the first, the technical team pushes hard without sufficiently translating the operational consequences. In the second, leadership waits too long because the issue has not been framed in a way that makes the need for a decision obvious. Neither pattern reflects strong incident management.

The National Cyber Incident Response Plan is also useful here because it distinguishes between different response roles and recognizes that cyber incidents often require coordination among multiple entities with distinct responsibilities.[25] That principle applies well beyond large national incidents. Even at smaller scales, infrastructure events frequently require the organization to think across technical response, asset response, continuity needs, regulatory concerns, and external coordination. The responder does not have to adopt a national-level structure to benefit from the underlying lesson. The lesson is that cyber incidents often outgrow simple technical ownership quickly, and organizations that do not prepare for that transition will struggle when it occurs.

A responder who operates well in this environment learns to answer a different set of questions than purely technical teams often emphasize. Instead of asking only, “What is happening in the network?” the responder must also ask, “Who must understand this now?” “What decision is approaching?” “What part of the organization owns that decision?” “What information is actually useful to them?” and “What happens if we delay this escalation by two hours, four hours, or one operational cycle?” These are leadership-support questions, but they remain fully within the discipline of

---

[25] Cybersecurity and Infrastructure Security Agency, *National Cyber Incident Response Plan* (Washington, DC: U.S. Department of Homeland Security, December 2016).

incident response because they determine whether technical findings translate into effective action.

For readers transitioning into critical infrastructure defense, this chapter marks another important adjustment. Being technically correct is not enough. In distributed environments, correctness must be made actionable. That requires more than expertise. It requires framing, restraint, role clarity, and an understanding that different stakeholders are not necessarily resisting the truth simply because they ask different questions. Often, they are responding to the same incident from different responsibility positions. The responder's task is to align those positions around a shared decision structure, not to pretend they do not exist.

This is what makes escalation a professional skill rather than an administrative step. Done poorly, escalation increases noise, spreads confusion, and creates the illusion of coordination without the reality of it. Done well, it turns uncertainty into a structured decision space where the organization can act before the incident outruns its ability to manage it.

Critical Thinking Question: What makes escalation effective in a complex organization, and how can a responder distinguish between merely informing leadership and actually enabling a decision?

Chapter 7: Rules of Engagement and Operational Restraint

Once an incident has been recognized, scoped at least provisionally, and elevated into the organizational decision space, responders often encounter a new source of friction. The problem is no longer only what the organization can do technically. The more important question becomes what the organization is authorized, prepared, and justified to do under the circumstances. This distinction matters because many technically possible actions carry legal, contractual, evidentiary, operational, or governance consequences that extend beyond the immediate cyber objective.

In practical terms, this means that technical feasibility and legitimate authority are not the same thing. A responder may identify an access path that appears to be enabling adversary activity, but the organization may not fully own that pathway, may depend on it for operational support, or may be obligated to coordinate with another party before altering it. A team may wish to isolate a system quickly, but that system may support a function whose interruption creates broader service, safety, or continuity consequences. A technically aggressive step may therefore be correct from a narrow security standpoint while still being wrong as an organizational decision.

This chapter is also where an important professional shift must be stated clearly. The transition described in this book is a transition from cyber warfighter to cyber defender. In civilian critical infrastructure environments, the responder's mission is not to pursue the attacker into external systems, retaliate against adversary infrastructure, or "destroy the enemy." The mission is to defend the organization's systems, reduce ongoing harm, preserve evidence, restore trustworthy operations, and coordinate with the appropriate internal leaders and external authorities.[26] In the United States, unauthorized access to other systems is generally governed by the Computer Fraud and Abuse Act, and the Department of Justice directs victims and private parties to report computer crimes to law enforcement rather than taking unauthorized action themselves.[27] That boundary is not a technical inconvenience. It is a defining feature of civilian cyber defense.

This is why incident response must operate inside clearly understood boundaries. The phrase "rules of engagement" is often associated with military operations, but the underlying idea is useful in civilian cyber defense as well. Every organization operates within a set of practical limits that define what actions may be taken, against which systems, under what conditions, and with what approvals.[28] Those limits are shaped by law, policy, ownership, contracts, regulation, evidence needs, service obligations, and internal authority structures. Whether an organization uses the phrase "rules of engagement" or not, those boundaries exist. Mature responders recognize them early and incorporate them into the way they frame recommendations.

This is especially important in critical infrastructure environments. The closer a cyber issue moves to operational technology, third-party

---

[26] Cybersecurity and Infrastructure Security Agency, *Federal Government Cybersecurity Incident and Vulnerability Response Playbooks* (Washington, DC: U.S. Department of Homeland Security, updated 2024), https://www.cisa.gov/resources-tools/resources/federal-government-cybersecurity-incident-and-vulnerability-response-playbooks.
[27] U.S. Department of Justice, *Justice Manual, 9-48.000 Computer Fraud and Abuse*, accessed May 20, 2026, https://www.justice.gov/jm/jm-9-48000-computer-fraud; U.S. Department of Justice, Computer Crime and Intellectual Property Section, "Reporting Computer, Internet-related, or Intellectual Property Crime," accessed May 20, 2026, https://www.justice.gov/criminal/criminal-ccips/reporting-computer-internet-related-or-intellectual-property-crime.
[28] Alexander Nelson, Sanjay Rekhi, Murugiah Souppaya, and Karen Scarfone, *Incident Response Recommendations and Considerations for Cybersecurity Risk Management: A CSF 2.0 Community Profile*, NIST SP 800-61 Rev. 3 (Gaithersburg, MD: National Institute of Standards and Technology, April 2025), https://doi.org/10.6028/NIST.SP.800-61r3.

support pathways, shared administrative systems, or public-facing services, the less likely it is that the cybersecurity team alone should make the most disruptive response decisions.[29] That does not weaken the cyber function. It reflects the reality that the consequences of action now extend beyond technical remediation. In these settings, the responder must understand that speed alone is not the measure of professionalism. Defensible action is.

Shared ownership is one of the most common reasons this issue becomes difficult. Many organizations rely on systems, services, and administrative pathways that are not neatly contained within one team's direct control. Vendor-managed tools, remote support mechanisms, managed services, outsourced monitoring, cloud administration, and federated access all create environments in which technical access and organizational authority do not line up perfectly. A cybersecurity team may be able to see the risk clearly, yet still lack the unilateral authority to alter the path along which that risk is moving. In other cases, the organization may own the system formally but still depend on a third party for safe modification, maintenance, or recovery. These are not edge cases. They are increasingly common operating conditions.

CISA's guidance on remote and network access security reflects part of this risk by emphasizing that external access paths, remote administration channels, and inherited trust relationships require stronger controls and closer management because they frequently become part of the attack surface.[30] The important lesson for responders is not simply that remote access can be dangerous. It is that the pathways through which

---

[29] Kevin Stouffer et al., *Guide to Operational Technology (OT) Security*, NIST SP 800-82 Rev. 3 (Gaithersburg, MD: National Institute of Standards and Technology, September 2023).

[30] Cybersecurity and Infrastructure Security Agency, *Modern Approaches to Network Access Security* (Washington, DC: U.S. Department of Homeland Security, June 18, 2024), https://www.cisa.gov/resources-tools/resources/modern-approaches-network-access-security.

organizations support operations may also be the pathways through which authority becomes complicated. When those paths are implicated in an incident, technical teams must be prepared to answer not only whether the path is risky, but also who owns it, what depends on it, and the operational cost of disrupting it.

Operational restraint becomes essential in that context. Restraint should not be misunderstood as passivity or fear. It is the professional discipline required to ensure that urgent action remains authorized, proportionate, and aligned to the mission. In many incidents, the instinctive technical recommendation is to shut down or sever the most suspicious path immediately. Sometimes that is exactly the right decision. In other cases, that action may cause avoidable harm if taken without understanding the surrounding dependencies, support implications, or evidentiary consequences. Restraint is what keeps the response team from confusing urgency with maturity.

Evidence preservation is one of the strongest examples of this principle. NIST's forensic guidance makes clear that organizations should integrate forensic thinking into incident response and should consult management and legal counsel regarding applicable laws, regulations, and handling requirements.[31] This matters because some actions that appear to reduce immediate risk can also destroy volatile evidence, complicate later reconstruction of the event, or undermine legal, contractual, or regulatory obligations. A hasty reboot, reconfiguration, or deletion may make the environment quieter, but it may also make the incident harder to understand, explain, or defend later. In environments where reporting, legal review, or external investigation may follow, that trade-off becomes significant.

---

[31] Karen Kent et al., *Guide to Integrating Forensic Techniques into Incident Response*, NIST SP 800-86 (Gaithersburg, MD: National Institute of Standards and Technology, September 2006).

This does not mean responders should preserve evidence at the expense of all other priorities. Cyber response is not a forensic museum exercise. In high-consequence incidents, the organization may need to act quickly to preserve continuity or reduce further harm. The point is that evidence should be treated as one of the consequences attached to the decision, not as an afterthought discovered after the fact. A professional responder should be able to say, in effect, that this action may reduce attacker opportunity, but it may also affect what evidence remains available, what reporting can be supported, and what the organization can later prove about what happened.

That same logic applies to operational consequences. NIST's OT security guidance emphasizes that operational environments often involve different tolerances for change, interruption, and safety-related impact than ordinary IT environments.[32] As incidents move closer to those environments, the responder must be able to explain not only the technical benefit of a proposed action but also the operational price. If isolating a system may impair monitoring, delay maintenance, interrupt continuity, or alter process behavior, those consequences must be visible to decision-makers. A recommendation that ignores them may still be technically sound, but it will not be operationally mature.

This is why the responder's role is not simply to demand action. It is to frame action responsibly. A useful recommendation should describe the technical risk, the likely consequence of inaction, the likely consequence of the proposed action, the authority or ownership issues attached to the system or path in question, and any additional coordination required before execution. That kind of framing does not slow the response unnecessarily. It improves the quality of the decision. It allows leadership,

---

[32] Kevin Stouffer et al., *Guide to Operational Technology (OT) Security*, NIST SP 800-82 Rev. 3 (Gaithersburg, MD: National Institute of Standards and Technology, September 2023).

operations, legal, and other stakeholders to evaluate the choice as a real organizational act rather than as a purely technical preference.

The same principle applies when the organization operates under policy or regulatory obligations that affect incident handling. Some incidents may trigger reporting requirements, contractual notice obligations, privacy concerns, or sector-specific coordination expectations. A responder does not need to become the organization's lawyer in order to recognize that such obligations exist. What the responder must do is avoid presenting technically aggressive action as though it were unconstrained by those realities. In mature environments, cybersecurity operates inside governance. It does not override governance simply because the risk is technical.

The line between cyber warfighting and civilian cyber defense is especially important here. In a military context, mission sets may include offensive authorities, intelligence support, or actions designed to impose cost on an adversary. In civilian critical infrastructure defense, those assumptions do not transfer cleanly. The defender's task is to protect the enterprise, support continuity, and enable lawful, defensible response. That means hardening, restricting, containing, preserving, recovering, and reporting. It does not mean crossing into unauthorized systems or improvising retaliation because the attacker appears to deserve it. The responder who fails to make that transition clearly is not being more aggressive or more mission focused. They are misunderstanding the mission.

This chapter is not an argument for caution at all costs. It is an argument for disciplined action within boundaries. Responders should be decisive. They should also be honest about what gives them the authority to act, what dependencies shape the outcome, and what forms of harm their recommendation may create or avoid. The strongest responders in critical infrastructure are not the ones who simply push for the fastest technical move. They are the ones who understand how to move the organization toward an action that is defensible in legal, operational, contractual, and mission terms.

That is the practical meaning of operational restraint. It protects the organization from self-inflicted damage at the very moment when urgency makes it more likely. It keeps technical teams from acting as though access equals authority. It forces consequence into the decision before consequence arrives through failure. Most importantly, it preserves the integrity of the response by ensuring that speed is never confused with justification.

The responder's task, then, is not merely to identify what can be done. It is to clarify what should be done, what may be done, what must be coordinated before action, and what risks remain if the organization delays. That is a more demanding standard than technical competence alone, but it is the standard that serious incident response requires.

Critical Thinking Question: When a technically sound action could also create legal, evidentiary, contractual, or operational problems, how should a responder present that option so the organization can act decisively without acting carelessly?

Part III. The Incident Response Lifecycle, Operationalized

The incident response lifecycle remains useful, but only when it is understood as a real-world operating model rather than a clean theoretical diagram. In practice, each phase carries friction, trade-offs, and consequences that are often underemphasized in formal descriptions. This section walks through preparation, detection, investigation, containment, eradication, and recovery as they are actually experienced in serious environments. The purpose of this part is to show how each stage of response depends not only on technical actions, but also on timing, organizational discipline, continuity concerns, and the ability to make defensible decisions under pressure.

Chapter 8: Preparation That Actually Works

Preparation is one of the most overclaimed ideas in cybersecurity. Most organizations say they are preparing. Many can point to plans, policies, diagrams, inventories, contact lists, and annual exercises as evidence that preparation exists. Some of those artifacts are useful and necessary. The problem is not that organizations create them. The problem is that they often confuse the existence of preparation materials with the existence of actual readiness. In practice, preparation matters only to the extent that it improves performance when the incident begins to apply pressure.

That standard is more demanding than most organizations prefer. It is relatively easy to document a process, assign responsibilities, and produce governance evidence that a response capability exists. It is much harder to build a capability that remains useful when the situation is unclear, key people are unavailable, logging is incomplete, support paths are slower than expected, and the organization must act before certainty is comfortable. NIST's current incident response guidance is explicit that incident response recommendations and considerations should be incorporated throughout cybersecurity risk management activities so organizations can prepare for incidents, reduce their number and impact, and improve detection, response, and recovery effectiveness.[33] That is a broader and more practical standard than plan ownership or annual review alone.

This is why preparation should be understood as a performance function, not a documentation function. A plan is not the goal. A plan is a tool. An inventory is not the goal. An inventory is a tool. A tabletop

---

[33] Alexander Nelson, Sanjay Rekhi, Murugiah Souppaya, and Karen Scarfone, *Incident Response Recommendations and Considerations for Cybersecurity Risk Management: A CSF 2.0 Community Profile*, NIST SP 800-61 Rev. 3 (Gaithersburg, MD: National Institute of Standards and Technology, April 2025), https://doi.org/10.6028/NIST.SP.800-61r3.

exercise is not the goal. It is a method for discovering whether the organization can still function when conditions begin to degrade. The real goal is organizational performance under stress. If the organization has many formal artifacts but still cannot align ownership, validate dependencies, reach the right people, or make decisions under pressure, then its preparation is weaker than it appears.

Meaningful readiness begins with understanding the environment well enough to defend it. That sounds obvious, but it is one of the first places organizations fail themselves. They know the systems they manage directly, but not always the systems they depend on. They know the major applications, but not always the administrative pathways that support them. They know the network boundaries in broad terms, but not always the trust relationships, service accounts, cloud dependencies, vendor access points, or operational sequences that matter most during disruption. In critical infrastructure environments, that weakness becomes especially costly because systems often matter not only because they exist, but because other functions depend on them. A smaller system with a narrow footprint may still become a major problem if it supports a continuity-critical process, a remote support function, or a safety-relevant dependency.[34]

For that reason, good preparation requires more than a technical asset inventory. It requires an understanding of ownership, identity, trust, and operational consequence. The organization needs to know which systems matter most, which identities matter most, what paths exist for remote administration, what systems rely on third-party support, what functions must remain available during disruption, and what dependencies could turn a local issue into a broader service problem. These are not

[34] Kevin Stouffer et al., *Guide to Operational Technology (OT) Security*, NIST SP 800-82 Rev. 3 (Gaithersburg, MD: National Institute of Standards and Technology, September 2023).

advanced maturity extras. They are the minimum conditions for making sound decisions once the environment becomes unstable.

Preparation also depends heavily on communication architecture. Many organizations assume they can improvise communication during an incident because the right people will naturally find each other once the problem becomes serious enough. Real incidents repeatedly disprove that assumption. Contact information becomes stale. Role ownership becomes uncertain. People assume someone else has already informed leadership, legal, operations, or outside support. Executive briefings vary in quality because no common structure exists. Technical teams provide detail without consequence framing. Leaders ask for certainty the responders cannot honestly provide. All of this is preventable only if the organization builds a communication structure before the incident instead of during it.

CISA's planning guidance reflects this point by treating incident response planning as a practical function that should establish roles, procedures, and communications before an event begins.[35] That does not mean every incident will unfold as the organization planned. It means the organization should not be discovering basic coordination mechanics for the first time while the incident is already creating pressure. A mature preparation model therefore includes updated escalation paths, clear decision rights, defined internal and external notification expectations, and a repeatable format for briefing technical, operational, and executive audiences.

Technical controls are equally subject to this same test of usability. Logging matters only if the relevant data can be accessed, interpreted, and trusted under incident conditions. Backups matter only if restoration has been validated in the way the environment will actually require.

---

[35] Cybersecurity and Infrastructure Security Agency, *Incident Response Plan (IRP) Basics* (Washington, DC: U.S. Department of Homeland Security, January 31, 2024), https://www.cisa.gov/resources-tools/resources/incident-response-plan-irp-basics.

Monitoring matters only if the team knows how to interpret the outputs in context rather than drowning in noise. Segmentation matters only if the organization understands what is really separated, what is still trusted, and what operational effects will follow if those boundaries must be enforced quickly. Preparation that stops at control deployment without validating operational use leaves the organization exposed to a different kind of failure: the false confidence that the tool exists, therefore the function is ready.

NIST's contingency planning guidance remains useful here because it emphasizes that contingency planning is tied to system availability, resilience, interrelationships with other emergency and security plans, and real-world recovery considerations rather than static documentation alone.[36] That view is important because cybersecurity incidents do not unfold separately from continuity. If the organization has never truly tested whether it can restore critical services, operate in a degraded condition, or sequence recovery in a disciplined way, then the existence of backup or recovery language in a document will not protect it during the event. The same is true of cyber event recovery planning more specifically. NIST's recovery guidance stresses planning, playbook development, testing, and continual improvement as part of recovery readiness, which reinforces the larger lesson that readiness must be exercised and refined rather than merely described.[37]

Third-party dependence is another area where superficial preparation often fails. Many organizations assume that because they have a vendor contract, support relationship, or managed service in place, that support will function as expected during an actual incident. That

---

[36] Marianne Swanson et al., *Contingency Planning Guide for Federal Information Systems*, NIST SP 800-34 Rev. 1 (Gaithersburg, MD: National Institute of Standards and Technology, May 2010).
[37] Michael Bartock et al., *Guide for Cybersecurity Event Recovery*, NIST SP 800-184 (Gaithersburg, MD: National Institute of Standards and Technology, December 2016).

assumption is often untested. The organization may not know how quickly the right vendor personnel can be reached, what the vendor is contractually obligated to do, what support is available after hours, what systems require vendor involvement before change is safe, or what happens if a vendor-administered pathway itself becomes part of the incident. In critical infrastructure environments, these questions are especially important because external support often touches systems or functions that cannot be altered casually. Preparation therefore requires more than vendor presence. It requires clarity about vendor roles, dependencies, escalation paths, and constraints before the relationship is stressed.

Exercises are supposed to expose these problems, but many exercises fail because they are designed to confirm confidence rather than reveal friction. A narrow tabletop that assumes full staffing, prompt communications, stable ownership, cooperative vendors, and perfect information may satisfy a schedule requirement without improving readiness meaningfully. A useful exercise introduces realistic inconvenience. It forces the organization to confront what happens when critical people are unavailable, when logs do not line up, when technical and operational perspectives diverge, or when a dependency turns out to matter more than anyone expected. The best exercises are not the ones that make the organization look polished. They are the ones that reveal what would fail if the event happened under less forgiving conditions.

This is especially important for readers moving into critical infrastructure defense from more centralized or technically contained environments. Preparation in infrastructure settings must account not only for cyber controls but also for operational consequences. It must recognize that some systems cannot be interrupted casually, that continuity may depend on people as much as on technology, that remote support may be part of both the operating model and the attack surface, and that incident response may unfold in the same timeline as weather events, public emergencies, supply disruption, or other non-cyber pressures. Preparedness in these settings is therefore inseparable from institutional realism. The question is not whether the organization has a plan. The question is whether the organization has prepared for the conditions under which the plan will actually be tested.

At its best, preparation preserves decision space. That is the deeper value. It does not remove uncertainty. It reduces avoidable uncertainty. It does not eliminate difficult choices. It prevents some of those choices from becoming harder than they needed to be. It does not guarantee a clean incident. It improves the organization's ability to respond coherently while the incident is still manageable. That is why meaningful preparation deserves a more serious standard than compliance completion. It should be judged by what it allows the organization to do when timing is poor, visibility is incomplete, and the cost of confusion is increasing.

Preparation matters because the incident should not be the first moment when the organization discovers what it does not understand about itself.

Critical Thinking Question: Which elements of preparedness are most likely to look sufficient during normal operations but fail quickly once an incident introduces ambiguity, degraded staffing, or operational pressure?

Chapter 9: Detection: Signal Versus Noise

Detection is often described as though it were mainly a tooling problem. Organizations are told to deploy monitoring technologies, collect telemetry, centralize logs, and build alerting logic that will identify suspicious activity quickly enough for the response team to act. Those steps matter, but they are only part of the problem. Monitoring can collect data without producing useful detection. Alerting can generate volume without generating clarity. Visibility can exist in the environment without being converted into understanding. For that reason, detection should be understood less as the mere presence of technical sensors and more as the organization's ability to recognize when weak or fragmented indicators have become operationally meaningful.

That distinction matters because most real environments produce far more noise than signal. Systems fail intermittently. Administrators connect at odd hours. Services restart unexpectedly. Users make mistakes. Applications generate noisy logs. Legacy systems behave inconsistently. Operational teams report irregular performance that may or may not have a cyber cause. Security tools themselves generate false positives, low-confidence warnings, and repeated events that quickly become familiar enough to ignore. In such an environment, the absence of perfect clarity is not unusual. It is normal. Detection quality therefore depends not on whether anomalies exist, but on whether the organization can interpret them with enough context to decide which ones matter.

NIST's guidance on continuous monitoring is useful here because it frames monitoring not as passive data accumulation, but as an ongoing program intended to provide visibility into assets, awareness of threats and vulnerabilities, and information that supports timely response to risk.[38] That

[38] Kelley Dempsey et al., Information Security Continuous Monitoring (ISCM) for Federal Information Systems and Organizations, NIST SP 800-137 (Gaithersburg, MD: National

is a much more demanding standard than simply collecting logs or deploying dashboards. It means that the organization must know what it is monitoring, why it is monitoring it, and how the resulting information will be interpreted when conditions become uncertain. Telemetry without interpretation is not detection. It is inventory.

This is one reason context matters so much. A single event rarely carries enough meaning by itself. The significance of an alert depends on where it occurred, when it occurred, who or what generated it, what system it touched, how normal or abnormal the activity is for that environment, and what the consequences would be if the event is part of a larger pattern. The same technical event may be low priority in one system and highly consequential in another. A valid remote login may be unremarkable in one part of the environment and deeply concerning in another. A repeated failed process may be routine instability in a low-consequence system and an early warning in a continuity-critical one. Detection becomes effective only when events are read inside operational context rather than in technical isolation.

The need for this context is one reason log management remains foundational to serious detection capability. NIST's log management guidance explains that logs are a critical source of information for understanding system and network activity, but it also makes clear that effective log management requires planning, standardization, storage, review, and analysis rather than simple collection.[39] Logs do not become useful merely by existing. They become useful when the organization can access them, correlate them, retain them appropriately, and understand how they support investigation and decision-making. A team that collects

---

Institute of Standards and Technology, September 2011), https://doi.org/10.6028/NIST.SP.800-137.

[39] Karen Kent and Murugiah Souppaya, Guide to Computer Security Log Management, NIST SP 800-92 (Gaithersburg, MD: National Institute of Standards and Technology, September 2006).

large amounts of logging data without clarity on ownership, relevance, retention, or analysis may still be weak at detection even if it looks mature from a tooling perspective.

This is one reason organizations often miss incidents they have, in a technical sense, already seen. The signal may have appeared in a monitoring console, a system log, an authentication record, an operational complaint, or a change event. The problem is that each fragment lived in a separate place, under separate ownership, or under a separate interpretation model. The authentication team saw unusual logins. Operations saw system irregularities. Security tooling produced low-confidence alerts. Administrators saw activity that looked strange but not necessarily hostile. No single event crossed the threshold for immediate action. The organization, therefore, concluded, often implicitly, that no incident was yet visible. In reality, the incident was visible in pieces.

This is where event correlation becomes more than a technical convenience. It becomes a defensive necessity. Detection improves when the organization can identify relationships among otherwise ordinary-looking events. A single abnormal access event may not be enough to justify concern. The same event paired with unusual timing, a maintenance mismatch, a system irregularity, and a sensitive administrative path may be very different. A low-confidence alert may not deserve immediate escalation in isolation, but it may become significant when it aligns with operator observations or access anomalies. The defender's task is not to react to every irregularity as though it were catastrophic. The defender's task is to recognize when multiple weak indicators are beginning to describe the same problem.

This is difficult because noisy environments train people to normalize inconsistency. That habit is not always unreasonable. If every minor irregularity triggered full incident procedures, organizations would exhaust themselves quickly. The challenge is that the same habit that protects teams from alert fatigue can also make them slow to recognize weak but dangerous patterns. Mature detection therefore requires more than sensitivity. It requires disciplined filtering. The organization must decide which systems matter most, which types of access deserve greater scrutiny, which anomalies are common enough to discount, and which combinations of conditions justify more serious attention.

CISA's guidance for organizations using logging on business and government systems reinforces this practical point by encouraging organizations not only to collect logs but to review key system logs and build the habit of using them operationally.[40] This is a useful reminder that detection is not purely a technology procurement function. It is also a practice. Teams become better at detection when they know what normal looks like, when they regularly review the kinds of indicators that matter in their environment, and when they are trained to connect technical and operational observations instead of leaving them in separate silos.

The challenge becomes even greater when the environment includes critical infrastructure or operational technology. In these settings, defenders may have less direct visibility, fewer easily deployable sensors, more legacy systems, and more non-cyber explanations for irregular behavior. A process anomaly may have a mechanical cause, a human cause, or a cyber cause. A remote connection may be part of ordinary vendor support or an abuse of legitimate access. A degraded operational response may be treated initially as a performance issue rather than a security issue. This does not make detection impossible. It makes context

---

[40] Cybersecurity and Infrastructure Security Agency, "Use Logging on Business Systems," accessed May 20, 2026, https://www.cisa.gov/audiences/small-and-medium-businesses/secure-your-business/use-logging-on-business-systems; Cybersecurity and Infrastructure Security Agency, "Use Logging on Government Systems," accessed May 20, 2026, https://www.cisa.gov/audiences/state-local-tribal-and-territorial-government/secure-us-sltt/use-logging-government-systems.

and collaboration more important. Cyber teams, operators, administrators, and engineers may each hold part of the truth. Detection fails when those parts are not brought together soon enough.

This is why detection should be treated as an organizational capability rather than a product feature. The organization needs visibility, but it also needs ownership clarity, baseline understanding, usable logging, review discipline, and a common habit of taking unexplained irregularities seriously enough to connect them before the evidence becomes overwhelming. NIST's continuous monitoring guidance supports this broader interpretation by linking visibility and awareness to timely risk response rather than to passive observation alone.[41] That is the right standard. Detection has value because it buys time. It shortens the period in which the adversary can act while the organization still believes it is only seeing background noise.

For students and newer practitioners, one of the most important lessons in this chapter is that serious incidents are often visible earlier than organizations realize. They are not always invisible. They are often misinterpreted. The problem is not that defenders never had access to the relevant evidence. The problem is that the evidence appeared in weak forms, in separate locations, or in combinations that no one had yet treated as meaningful. That is why the discipline of detection involves more than technical review. It involves the judgment to ask whether separate irregularities might belong to the same emerging event.

Good detection does not eliminate ambiguity. It reduces how long the organization remains trapped inside it. That is a more practical and more realistic standard than the expectation that early detection will

---

[41] Kelley Dempsey et al., Information Security Continuous Monitoring (ISCM) for Federal Information Systems and Organizations, NIST SP 800-137 (Gaithersburg, MD: National Institute of Standards and Technology, September 2011), https://doi.org/10.6028/NIST.SP.800-137.

produce certainty. The responder's job is not to turn every weak signal into a declaration. It is to recognize when weak signals have accumulated enough context that inaction becomes a greater risk than disciplined escalation.

Critical Thinking Question: What determines whether a weak indicator remains ordinary background noise or becomes an early signal of a meaningful incident, and how should organizations train themselves to recognize the difference?

Chapter 10: Investigation and Scoping Under Pressure

Once an organization believes it may be facing a real incident, the next challenge is determining what that incident actually touches. That sounds straightforward, but it rarely is. Scope is not simply a list of affected systems. It is the organization's working understanding of which identities, systems, services, access paths, and dependencies are implicated, which ones are at risk, and which parts of the picture remain unresolved. In practice, that understanding is built gradually, revised repeatedly, and shaped by the fact that the incident continues to evolve while the investigation is still underway.

This is why scoping should not be treated as a one-time discovery. Organizations often want a hard boundary early because leadership needs decisions, operations needs continuity guidance, and responders need to prioritize. Those pressures are real, but they can push teams toward false confidence. In the early stages of a serious incident, the more realistic goal is not to define the perfect boundary of the event. The goal is to improve the quality of the next decision by expanding understanding faster than the incident expands consequence.

NIST's incident response guidance is useful here because it treats analysis as an ongoing part of incident handling rather than a discrete task completed before response decisions.[42] That matters because scoping and response usually proceed together. The organization does not get to pause the incident while it performs a complete technical study. It must often contain, coordinate, and communicate while still learning what the incident actually is. A responder who assumes scope must be fully settled

---

[42] Alexander Nelson, Sanjay Rekhi, Murugiah Souppaya, and Karen Scarfone, Incident Response Recommendations and Considerations for Cybersecurity Risk Management: A CSF 2.0 Community Profile, NIST SP 800-61 Rev. 3 (Gaithersburg, MD: National Institute of Standards and Technology, April 2025), https://doi.org/10.6028/NIST.SP.800-61r3.

before meaningful action can occur will often lose time the organization cannot afford.

The first questions in scoping should therefore be disciplined and consequence-oriented. What is confirmed? What is only suspected? Which identities appear exposed? Which systems show related activity? What access paths connect the affected environment to adjacent systems or services? What dependencies could turn a local issue into a broader continuity problem? Which unknowns could materially change the next decision if they turn out to be true? These questions matter because they help distinguish between technical curiosity and operationally relevant investigation. Not every detail needs to be known immediately. The right details do.

This is one reason hypothesis-driven investigation is so important. Responders need a working theory of the incident in order to know where to look next, but that theory must remain provisional. If the first explanation becomes emotionally or organizationally comfortable, teams can begin treating it as settled before the evidence supports that conclusion. That is one of the most common scoping failures in incident response. The organization does not lack intelligence or effort. It becomes too attached to the first theory that seems coherent and then investigates in ways that confirm that theory instead of testing it.

A better approach is to build a working explanation and then actively look for evidence that would weaken or expand it. If the team believes the issue is limited to one system, what evidence would show that the issue is actually identity-based and broader? If the team believes the event is tied to a single access path, what evidence would suggest multiple points of persistence or movement? If the team believes the impact is local, what dependencies would prove that the effect is already crossing into adjacent systems or services? This is not skepticism for its own sake. It is disciplined investigation intended to prevent narrow assumptions from hardening into operational mistakes.

Identity plays a major role in this phase. Modern incidents often expand through accounts, trust relationships, service credentials, remote administration paths, or federated access rather than through obviously visible malware movement alone. That reality changes scoping significantly. A responder who focuses only on the first visibly affected device may miss the more important issue of which accounts were used, what privileges they carried, what systems recognized those identities, and what trust relationships made the activity possible. In many cases, the scope of an incident is larger in identity terms than in host terms. The compromised server or endpoint may be only the first place the organization noticed the problem, not the best indicator of where the problem actually extends.

This is also why forensic thinking must be integrated into scoping rather than treated as a separate specialty that begins later. NIST's forensic guidance emphasizes that incident response should incorporate forensic techniques so the organization can collect, examine, and analyze information in ways that support understanding and preserve useful evidence.[43] Scoping depends on this discipline because early investigative actions can either increase understanding or unintentionally narrow it. If teams fail to preserve relevant data, overwrite volatile evidence, or alter affected systems before understanding what questions matter most, they may reduce their own ability to determine how far the incident extends.

At the same time, scoping must be realistic about visibility limits. Not every environment provides complete logging, full telemetry, or immediate ownership clarity. Continuous monitoring guidance is helpful

---

[43] Karen Kent, Suzanne Chevalier, Tim Grance, and Hung Dang, Guide to Integrating Forensic Techniques into Incident Response, NIST SP 800-86 (Gaithersburg, MD: National Institute of Standards and Technology, September 2006).

here because it frames visibility as a risk management function rather than as a guarantee of certainty.[44] Responders must often work with partial evidence, uneven coverage, or delayed access to the systems that matter most. The right response to that condition is not paralysis. It is disciplined uncertainty management. The team should state clearly what it knows, what it believes, what it cannot yet validate, and how those unknowns affect the next decision.

This is especially important in critical infrastructure and mixed IT/OT environments. In these settings, complete scoping may be even harder because telemetry is inconsistent, asset ownership may cross functional boundaries, and process irregularities may have cyber and non-cyber explanations. A system anomaly may reflect technical compromise, process instability, mechanical failure, or some combination of these. An access event may look valid from a network perspective while still being abnormal from an operational perspective. A responder who relies only on technical visibility may misread a broader operational problem. Conversely, a responder who treats every operational irregularity as a cyber indicator may distort the picture in the other direction. Good scoping therefore depends on multiple perspectives being brought into the same analytical process before the organization commits too strongly to a narrow interpretation.

This is where communication becomes part of investigation. Scoping is not useful if it remains trapped inside technical notes that no one else can act on. Leadership, operations, legal, and external partners may all need some understanding of the current scope, but they do not all need the same detail. What they need is an honest description of what is

---

[44] Kelley Dempsey et al., Information Security Continuous Monitoring (ISCM) for Federal Information Systems and Organizations, NIST SP 800-137 (Gaithersburg, MD: National Institute of Standards and Technology, September 2011), https://doi.org/10.6028/NIST.SP.800-137.

confirmed, what appears likely, what remains unresolved, and what those uncertainties mean for the next decision window. A responder who claims too much will damage trust later when the scope expands. A responder who refuses to say anything until the scope is perfect will deny the organization the chance to act while options still exist. The discipline lies in speaking clearly without pretending certainty.

The CISA federal playbooks are useful here because they treat investigation, evidence development, and response action as interdependent rather than sequentially isolated.[45] That is the correct mindset for serious incidents. Scoping is not an academic exercise performed for completeness. It is a decision-enabling function. The reason to improve scope is not simply to know more. It is to decide better. Better scope informs better containment. Better scope informs better communications. Better scope informs better continuity planning. Better scope informs better judgment about whether the issue is local, expanding, or already beyond the organization's first assumptions.

That last point is worth emphasizing. One of the most dangerous moments in incident response is the point at which the incident stops getting smaller. Organizations often begin an investigation hoping the problem will collapse into a familiar and manageable explanation. Sometimes it does. When it does not, the team must be willing to let the scope expand honestly rather than resist that expansion because it complicates operations, communications, or leadership comfort. The cost of acknowledging a broader scope may be high. The cost of denying it is usually higher.

---

[45] Cybersecurity and Infrastructure Security Agency, Federal Government Cybersecurity Incident and Vulnerability Response Playbooks (Washington, DC: U.S. Department of Homeland Security, updated 2024), https://www.cisa.gov/resources-tools/resources/federal-government-cybersecurity-incident-and-vulnerability-response-playbooks.

Good scoping does not promise certainty on demand. It promises disciplined improvement in understanding under pressure. That is a more realistic and more useful standard. The responder's responsibility is not to produce a perfect technical map before the organization can move. The responsibility is to ensure that each round of investigation informs the next decision, makes it more proportional, and reduces its dependence on unsupported assumptions.

Critical Thinking Question: Why is it dangerous to define incident scope too early, and how should responders communicate evolving scope in a way that supports action without creating false confidence?

Chapter 11: Containment Is a Business Decision

Containment is often portrayed as a purely technical phase of incident response. In simplified models, the idea appears straightforward. The organization identifies the malicious activity, isolates the affected systems, blocks the attacker's access, and limits further spread. That description is directionally correct but incomplete. In real incidents, especially in critical infrastructure environments, containment is one of the most difficult decision points in the entire lifecycle because it forces the organization to choose which risks it will accept immediately in order to prevent worse risks later.

That is why containment should not be understood as a simple synonym for shutdown. Its actual purpose is to limit further harm while preserving the organization's ability to continue operating, investigate, recover, and make defensible decisions.[46] In some cases, full isolation may be the right answer. In other cases, full isolation may create operational, safety, continuity, or public-consequence effects that are more damaging than a narrower and more disciplined containment step. The responder's task is therefore not merely to identify what could be disconnected. The task is to help the organization decide what must be restricted, what can remain online temporarily, and what consequences attach to either choice.

This becomes especially important in critical infrastructure environments because the systems involved often support processes that cannot be interrupted casually.[47] A system may be technically suspect and

---

[46] Cybersecurity and Infrastructure Security Agency. Federal government cybersecurity incident and vulnerability response playbooks. U.S. Department of Homeland Security, updated 2024, https://www.cisa.gov/resources-tools/resources/federal-government-cybersecurity-incident-and-vulnerability-response-playbooks.
[47] Kevin Stouffer et al., Guide to Operational Technology (OT) Security, NIST SP 800-82 Rev. 3 (Gaithersburg, MD: National Institute of Standards and Technology, September 2023).

still support treatment continuity, fuel movement, logistics coordination, safety monitoring, or another essential function. In these environments, the question is rarely, "Can we shut this down?" The more useful question is, "What happens if we do?" A containment step that makes perfect sense inside a narrow cybersecurity model may still be the wrong first move if it blinds operators, interrupts a continuity-critical process, prevents safe maintenance, or creates new operational instability before the organization has a viable fallback.

This is one reason containment is better understood as a business and mission decision informed by cybersecurity rather than as a cybersecurity decision imposed on the rest of the organization. The technical team is central to identifying the risk, explaining the access path, and estimating how the threat may continue to operate if nothing changes. But the consequences of containment often extend into continuity, operations, legal exposure, contractual obligations, safety, and public trust. Once the choice reaches that level, leadership and operational owners must be part of the decision. The cyber team's responsibility is not reduced by this. It becomes more demanding. The team must now frame the recommendation in terms of both technical effect and operational consequence.

CISA's federal incident response playbooks are useful on this point because they treat containment as part of a coordinated response process rather than as an isolated technical action.[48] That framing matters. Containment usually has second-order effects. Restricting an account may disrupt legitimate administration. Segmenting a network may interrupt process dependencies. Blocking a support pathway may reduce attacker

---

[48] Cybersecurity and Infrastructure Security Agency. Federal government cybersecurity incident and vulnerability response playbooks. U.S. Department of Homeland Security, updated 2024, https://www.cisa.gov/resources-tools/resources/federal-government-cybersecurity-incident-and-vulnerability-response-playbooks.

opportunity while also delaying restoration or vendor assistance. Requiring reauthentication may improve security while also slowing operations at a moment when continuity is already under strain. A recommendation that ignores these effects may still sound technically strong, but it does not yet qualify as decision-grade advice.

This is why responders should distinguish between short-term containment and sustainable containment. Short-term containment buys time. It reduces the attacker's ability to continue moving while the organization clarifies scope, confirms dependencies, and decides what broader actions are possible. Sustainable containment is different. It is intended to stabilize the environment in a way that the organization can maintain while investigation, eradication, and recovery continue. These two forms of containment are related, but they are not identical. A short-term restriction that is acceptable for thirty minutes may be harmful if sustained for three days. A rapid isolation decision that is tolerable during immediate triage may become operationally unacceptable once continuity obligations begin to accumulate.

That distinction is especially useful in operational technology and critical infrastructure settings. NIST's OT security guidance emphasizes that these environments have unique performance, reliability, and safety requirements.[49] That means containment must often be phased, selective, or conditional. A responder may recommend increased monitoring around one segment, restricted remote access through another, manual fallback for one process, and delayed isolation of a different dependency until operators are prepared. To someone outside critical infrastructure, that may appear indecisive. In reality, it may be the most disciplined and

---

[49] Kevin Stouffer et al., Guide to Operational Technology (OT) Security, NIST SP 800-82 Rev. 3 (Gaithersburg, MD: National Institute of Standards and Technology, September 2023).

responsible available path because it reduces attacker opportunity without creating immediate process instability or service loss.

This is where consequence mapping becomes central. Before recommending containment, responders should be able to explain what the action is expected to accomplish and what it is expected to cost. If an account is disabled, what administrative or operational work will stop with it? If a network path is blocked, what trusted function may also be disrupted? If a system is isolated, what telemetry, process visibility, or control capability may be lost? If nothing is done immediately, what access or movement does the attacker likely retain? These are not side questions. They are the actual decision space. Without them, containment becomes little more than technical preference under stress.

CISA's Eviction Strategies Tool is helpful in illustrating this broader reality because it is designed to match observed adversary activity with recommended response actions for containment and eviction.[50] The tool's value is not that it produces an automatic answer. Its value is that it reinforces a disciplined connection between what the adversary is doing and what response options may disrupt that activity. That is exactly how containment should be approached. The organization should not ask only, "What can we shut down?" It should ask, "What attacker activity are we trying to interrupt, what paths support that activity, and what organizational price will follow from each available option?"

This is also why partial containment should not be treated as weak containment by default. In some environments, a selective restriction is the most responsible immediate action available. If the organization can limit remote access, tighten identity controls, increase monitoring, reduce trust relationships, or move selected functions into a degraded but stable state

---

[50] Cybersecurity and Infrastructure Security Agency. Eviction Strategies Tool. U.S. Department of Homeland Security, July 30, 2025, https://www.cisa.gov/resources-tools/resources/eviction-strategies-tool.

while preserving essential services, that may be preferable to a sweeping technical move that creates avoidable operational harm. The question is not whether the first action looks dramatic. The question is whether it meaningfully improves the organization's position without causing disproportionate damage.

Of course, partial containment can also become an excuse for hesitation if it is not tied to a broader plan. That is why responders must be careful not to confuse incremental action with sufficient action. Every containment choice should be evaluated against the same two standards. First, does it materially reduce attacker opportunity? Second, is the organization prepared to absorb the consequence of doing more if that becomes necessary? A selective action that neither meaningfully constrains the adversary nor preserves options for follow-on containment is not disciplined. It is simply smaller.

Continuity planning becomes inseparable from this decision. NIST's contingency planning guidance remains relevant because it emphasizes the interrelationship between continuity, recovery, and operational resilience.[51] Containment decisions made without reference to continuity assumptions are often more dangerous than they first appear. If the organization does not know which services must remain available, what degraded modes are possible, what manual workarounds exist, and what dependencies matter most in the next operational window, then it is not really deciding on containment. It is gambling on disruption. A mature responder should therefore think about containment and continuity together, not as separate conversations held by separate teams.

This is particularly important for readers making the shift from cyber warfighter to cyber defender. In more centralized or offensive-minded

---

[51] Marianne Swanson et al., Contingency Planning Guide for Federal Information Systems, NIST SP 800-34 Rev. 1 (Gaithersburg, MD: National Institute of Standards and Technology, May 2010).

contexts, there can be a strong instinct to remove adversary access as aggressively and quickly as possible. In civilian critical infrastructure, that instinct must be disciplined by consequence. The defender's mission is not simply to deny the attacker. It is to protect the enterprise, preserve essential function, and enable lawful and sustainable recovery. That means the best containment action is not always the fastest one in narrow technical terms. It is the one that imposes acceptable immediate cost in order to prevent greater harm later.

The leadership challenge here is straightforward even if the decision is not. Leaders need recommendations that are technically informed, operationally framed, and explicit about trade-offs. They do not need a false promise that containment can be achieved without cost. In most serious incidents, there will be cost. The relevant question is which cost the organization is better positioned to absorb now. A responder who can explain that clearly is doing more than reporting technical findings. The responder is helping leadership choose how much disruption to accept in order to avoid a worse loss of control.

Containment is therefore one of the clearest points at which cyber response stops being a purely technical function and becomes an exercise in risk governance under pressure. The organization is not deciding whether the attacker is bad. That is already established. It is deciding what it is willing to interrupt, isolate, delay, or degrade in order to keep the situation from becoming harder to recover from. That is why containment belongs among the most consequential decisions in incident response. It is not just about stopping adversary movement. It is about deciding which pain is more survivable.

Critical Thinking Question: When every realistic containment option imposes some form of immediate cost, what makes one option more responsible than another, and who should decide where that line is drawn?

Chapter 12: Eradication and the Persistence Problem

Containment and eradication are often discussed as though they are adjacent steps in a neat sequence. In practice, they are related but fundamentally different. Containment is designed to limit immediate harm, restrict attacker movement, and preserve the organization's ability to function while the incident is still unfolding. Eradication is more demanding. It requires the organization to remove the attacker's ability to continue operating, return later, or maintain influence through identities, services, trust relationships, administrative pathways, or other forms of persistence.[52] A team can contain visible malicious activity without truly eliminating the conditions that made that activity possible.

This distinction matters because many organizations declare success too early. Once the most obvious signs of malicious activity have quieted, there is a natural institutional desire to believe the incident is now under control. The systems appear calmer. Alerts diminish. Users stop reporting obvious symptoms. Leadership begins asking whether the issue has been resolved. Those signals may reflect real progress, but they do not prove eradication. They may only prove that the adversary has lost one pathway, paused activity temporarily, or shifted into a less visible posture. A quieter environment is not necessarily a clean environment.

This is why eradication should not be confused with cleanup. Deleting malicious files, rebuilding a host, changing a password, or disabling one suspicious connection may all be useful actions. None of them, standing alone, answers the deeper question of whether the attacker's access has actually been removed. An organization that focuses only on the most visible artifacts of the incident can end up remediating

---

[52] Alexander Nelson, Sanjay Rekhi, Murugiah Souppaya, and Karen Scarfone, Incident Response Recommendations and Considerations for Cybersecurity Risk Management: A CSF 2.0 Community Profile, NIST SP 800-61 Rev. 3 (Gaithersburg, MD: National Institute of Standards and Technology, April 2025), https://doi.org/10.6028/NIST.SP.800-61r3.

symptoms while leaving persistence mechanisms intact. The result is often a false sense of closure followed by renewed compromise, delayed reinfection, or the discovery that the adversary's most important foothold was never the first one the defenders saw.

Modern incidents make this especially difficult because persistence is no longer primarily a malware problem. It is increasingly an identity and trust problem.[53] If a privileged account remains exposed, if a service credential remains valid, if a remote administration channel remains trusted, or if a federated identity relationship remains abusable, then the attacker may not need the original malicious toolset to reestablish influence. This is one reason zero trust thinking is so relevant to eradication. NIST's zero trust guidance emphasizes that security decisions should focus on users, assets, and resources rather than on broad assumptions of internal trust.[54] In eradication terms, that means the organization must ask not only what code or artifacts were present, but what trust was abused and whether that trust has actually been rebuilt on a justified basis.

This shift has practical consequences for responders. A rebuilt server does not guarantee eradication if the same administrative identities still provide the attacker a path back into the environment. A cleaned endpoint does not guarantee eradication if the same support mechanism can be reused. A changed password does not guarantee eradication if the underlying privilege model remains weak, if related credentials were also exposed, or if the attacker still retains access through another trusted path. The organization must therefore think at the level of access and architecture, not only at the level of devices and files.

---

[53] Cybersecurity and Infrastructure Security Agency, Eviction Strategies Tool (Washington, DC: U.S. Department of Homeland Security, July 30, 2025), https://www.cisa.gov/resources-tools/resources/eviction-strategies-tool.
[54] Scott Rose et al., Zero Trust Architecture, NIST SP 800-207 (Gaithersburg, MD: National Institute of Standards and Technology, August 2020).

CISA's Eviction Strategies Tool reflects this reality by focusing on adversary behavior and response actions intended to contain and evict threat actors based on observed tactics, techniques, and procedures.[55] Its value lies in the reminder that eradication is not simply a matter of "cleaning" the environment. It is a matter of deliberately disrupting the methods by which the adversary retains presence and control. The organization must know what it is trying to evict, what paths enable that activity, what controls can disrupt it, and what sequencing is required to do so without leaving open routes for immediate return.

Sequencing is especially important. In some incidents, defenders are tempted to reset credentials, reimage systems, or restore services in whatever order seems fastest. That approach can create new weakness if the sequence leaves trust relationships open or restores compromised connections before the organization understands how they were abused. Eradication often requires a more deliberate progression. Accounts may need to be reviewed, prioritized, and reset in coordination with service restoration. Administrative access may need to be constrained before systems are rebuilt. Monitoring may need to increase before a major reset so the organization can recognize whether the attacker is attempting to adapt. Restoration may need to wait until the most consequential trust paths have been hardened. This is not bureaucracy. It is the practical discipline required to keep the organization from restoring the same weakness it is trying to remove.

The challenge becomes even greater in operational technology and critical infrastructure environments. NIST's OT security guidance makes clear that these environments have unique performance, reliability, and safety requirements that constrain how quickly or aggressively systems

---

[55] Cybersecurity and Infrastructure Security Agency, Eviction Strategies Tool (Washington, DC: U.S. Department of Homeland Security, July 30, 2025), https://www.cisa.gov/resources-tools/resources/eviction-strategies-tool.

can be altered.[56] A broad reimage strategy that may be viable in parts of enterprise IT may be operationally unrealistic in process-dependent environments. Some systems cannot be restarted casually. Some paths cannot be revoked instantly without affecting continuity or process visibility. Some dependencies involve vendors, support arrangements, or industrial protocols that make rapid eradication slower and more constrained than defenders would prefer. In these settings, eradication is often a phased effort rather than a single decisive event.

This reality creates pressure on the responder to communicate honestly about confidence. Leadership may want a simple answer to the question of whether the threat is gone. The technically honest answer may be more conditional. The organization may be able to say that known malicious activity has been interrupted, that the most exposed accounts have been reset, that remote access has been restricted, and that key paths of persistence have been addressed. It may not yet be able to say with certainty that every foothold has been removed. That is not weakness. It is disciplined honesty. Overstating confidence in eradication can be more harmful than admitting that the organization has reduced risk significantly but is still validating whether the environment can be trusted again.

This is where many organizations become vulnerable to institutional impatience. After the visible crisis begins to calm, the pressure to move toward recovery grows. Business leaders want timelines. Operators want stable conditions. Executives want assurance. External stakeholders may want confirmation that the problem is over. Those demands are understandable, but they can lead the organization to compress the eradication phase into something more symbolic than real. A team that

---

[56] Kevin Stouffer et al., Guide to Operational Technology (OT) Security, NIST SP 800-82 Rev. 3 (Gaithersburg, MD: National Institute of Standards and Technology, September 2023).

has not yet rebuilt trust in identities, access paths, and dependencies may be pushed into calling the problem solved because the organization is exhausted by uncertainty. That is one of the most dangerous moments in an incident lifecycle.

A more mature approach treats eradication as a trust reconstruction problem. The organization must be able to explain why it now believes the environment is safer than it was before. Which identities were reviewed and reset? Which access relationships were restricted or redesigned? Which pathways were eliminated, monitored, or reduced in privilege? Which services were restored only after validation rather than by assumption? Which parts of the environment remain under enhanced scrutiny because certainty is still building? These are the kinds of questions that distinguish technical cleanup from meaningful eradication.

NIST's incident response guidance supports this broader framing because it treats response and recovery as activities integrated into cybersecurity risk management rather than as isolated technical events.[57] Eradication is part of that larger management problem. It requires the organization to think in terms of attacker opportunity, trust abuse, and residual risk rather than only in terms of visible compromise. The responder's job is not to create the appearance of finality. The responder's job is to help the organization remove adversary access in a way that is operationally defensible and technically honest.

This is especially important for professionals transitioning from a cyber warfighter mindset into a cyber defender role. In more offensive or highly centralized contexts, it may be tempting to imagine eradication primarily as the decisive elimination of the adversary's presence. In civilian

---

[57] Alexander Nelson, Sanjay Rekhi, Murugiah Souppaya, and Karen Scarfone, Incident Response Recommendations and Considerations for Cybersecurity Risk Management: A CSF 2.0 Community Profile, NIST SP 800-61 Rev. 3 (Gaithersburg, MD: National Institute of Standards and Technology, April 2025), https://doi.org/10.6028/NIST.SP.800-61r3.

critical infrastructure, the mission is different. The mission is to restore trustworthy operation through disciplined reduction of attacker access, careful management of dependencies, lawful action, and defensible sequencing. The defender is not “destroying the enemy.” The defender is rebuilding the conditions under which the enterprise can operate without granting the adversary continued opportunity. That difference in mission should shape the language and logic of the entire eradication phase.

A useful way to think about eradication, then, is this: if the organization cannot explain why it trusts the environment again, it has not finished the work. Silence in the logs is not enough. The disappearance of obviously malicious activity is not enough. A successful rebuild of one system is not enough. Eradication is complete only when the organization has removed, reduced, or controlled the paths through which the adversary was able to persist and when it can justify that judgment with evidence rather than hope.

Critical Thinking Question: If visible malicious activity has stopped but the organization still cannot explain why it should trust the environment again, has eradication actually occurred, or has the incident only become quieter?

Chapter 13: Recovery Means Restoring Trust

Recovery is often misunderstood as the moment when systems come back online. That view is too narrow. Technical restoration is part of recovery, but it is not the same thing as recovery itself. A system can be reachable, responsive, and apparently functional while the organization still lacks a defensible basis for trusting it. NIST's recovery guidance is explicit that organizations improve resilience when recovery planning is comprehensive, resources are identified and prioritized, and plans are tested and improved over time.[58] That framework points to a broader truth: recovery is not merely the reactivation of technology. It is the disciplined restoration of trustworthy operations.

This distinction matters because incidents create strong pressure to resume service as quickly as possible. Operations wants normal function restored. Leadership wants stability. Stakeholders want reassurance. In critical infrastructure environments, that pressure is even more intense because the systems involved often support essential services rather than only internal business processes. When the organization is under visible strain, the desire to reconnect systems and resume activity can quickly outrun the evidence needed to justify that decision. Recovery fails when speed becomes the only metric.

A more useful way to think about recovery is to separate availability from confidence. Availability answers whether a system can be used. Confidence answers whether it should be trusted in its current state. Those are not the same question. A restored server may still rely on identities, trust relationships, administrative pathways, or dependencies that were implicated during the incident. A resumed service may still be operating under incomplete validation. A network segment may be active

---

[58] Michael Bartock, Jeffrey Cichonski, Murugiah Souppaya, Matthew Smith, Gregory Witte, and Karen Scarfone, Guide for Cybersecurity Event Recovery, NIST SP 800-184 (Gaithersburg, MD: National Institute of Standards and Technology, December 2016).

again even though the organization has not yet resolved whether monitoring, segmentation, credential integrity, and supporting controls are strong enough to support normal reliance. Recovery becomes dangerous when the organization treats technical responsiveness as proof of restored trust.

That is why validation is central to the recovery phase. Before a system, service, or process is brought back into normal operation, the organization should be able to explain what has been verified, what remains under heightened scrutiny, and what temporary controls remain in place to manage residual risk.[59] In some environments, that validation may include confirming that compromised identities have been reset, that privileged access has been reviewed, that administrative paths have been restricted appropriately, that configuration integrity has been checked, that logging is functioning, and that monitoring has been tuned to detect signs of recurring abuse. Recovery is stronger when those steps are explicit rather than assumed.

Service prioritization is also critical. NIST's contingency planning guidance emphasizes the need to identify and prioritize resources and functions as part of preparedness and recovery planning.[60] That principle carries directly into incident recovery. Organizations should not restore systems in whatever order is most convenient or most visible. They should restore in the order that best supports mission, continuity, and safe operation. In a critical infrastructure context, some supporting systems may need to be trusted first because they enable safe monitoring, process

---

[59] Alexander Nelson, Sanjay Rekhi, Murugiah Souppaya, and Karen Scarfone, Incident Response Recommendations and Considerations for Cybersecurity Risk Management: A CSF 2.0 Community Profile, NIST SP 800-61 Rev. 3 (Gaithersburg, MD: National Institute of Standards and Technology, April 2025), https://doi.org/10.6028/NIST.SP.800-61r3.
[60] Marianne Swanson, Pauline Bowen, Amy Phillips, Dean Gallup, and David Lynes, Contingency Planning Guide for Federal Information Systems, NIST SP 800-34 Rev. 1 (Gaithersburg, MD: National Institute of Standards and Technology, May 2010; updated November 11, 2010).

control, or continuity decisions for more visible services. A poor recovery sequence can create the impression of progress while actually making the environment less stable.

This is particularly important in operational technology and infrastructure settings. NIST's OT security guidance makes clear that these environments include unique performance, reliability, and safety requirements.[61] That means recovery may have to proceed more deliberately than in ordinary enterprise IT. The organization may need to keep certain systems in degraded mode longer than leadership would prefer. Manual workarounds may remain necessary while trust is rebuilt. Some connections may need to stay restricted even after services resume. In these environments, "normal operations resumed" can be a misleading phrase if it obscures the fact that recovery is still being managed under temporary controls and heightened caution.

A disciplined recovery posture, therefore, accepts that degraded operations may be a sign of maturity rather than a sign of failure. When an organization uses temporary restrictions, heightened review, staged restoration, or manual fallback procedures, it is not necessarily demonstrating weakness. It may be demonstrating that it understands the difference between immediate pressure and justified trust. The purpose of those temporary conditions is not to delay recovery for its own sake. The purpose is to avoid rebuilding the same exposure that the incident just revealed.

This is where communication becomes especially important. Leadership and stakeholders need more than reassurance. They need honest recovery framing. That means distinguishing among what has been restored, what has been validated, what remains restricted, and what

---

[61] Kevin Stouffer et al., Guide to Operational Technology (OT) Security, NIST SP 800-82 Rev. 3 (Gaithersburg, MD: National Institute of Standards and Technology, September 2023).

residual risks are still being managed.[62] Overstating recovery can damage trust later if the organization has to reverse course, re-isolate a system, or explain why a restored function was not as secure as earlier statements implied. Careful language is not bureaucratic caution. It is part of the technical integrity of the recovery effort.

The same principle applies to confidence statements. The organization should avoid binary language where the truth is more conditional. It is often more accurate to say that core services have been restored under enhanced monitoring, that selected systems remain under restriction pending further validation, or that key functions have resumed while trust in supporting dependencies is still being rebuilt. These distinctions matter because they reflect the actual state of risk. Recovery should not be narrated as a symbolic return to normal if the environment is still operating under temporary assumptions, constrained trust, or incomplete validation.

For readers moving into critical infrastructure defense from more centralized or offense-oriented environments, this chapter reinforces another important transition. The civilian defender's mission is not simply to reestablish operation as quickly as possible. It is to restore operation on a defensible basis. That means the standard for recovery is not speed alone. It is whether the organization can explain why it now believes the environment is safe enough, stable enough, and trustworthy enough to support the services people depend on. That is a more demanding standard, but it is the right one.

This is also why recovery should be viewed as a bridge to resilience rather than as the last page of the incident. NIST's event recovery

---

[62] Alexander Nelson, Sanjay Rekhi, Murugiah Souppaya, and Karen Scarfone, Incident Response Recommendations and Considerations for Cybersecurity Risk Management: A CSF 2.0 Community Profile, NIST SP 800-61 Rev. 3 (Gaithersburg, MD: National Institute of Standards and Technology, April 2025), https://doi.org/10.6028/NIST.SP.800-61r3.

guidance ties planning, testing, prioritization, and lessons learned to organizational resilience precisely because recovery decisions shape whether the next incident begins in a stronger or weaker environment.[63] If the organization restores services without improving trust boundaries, identity discipline, monitoring visibility, or dependency awareness, then recovery may only return it to the same weakness that made the incident costly in the first place. Real recovery therefore includes both restoration and correction.

In the end, recovery is complete not when the organization is tired of the incident, not when executives want closure, and not when the most visible systems are available again. Recovery is complete when the organization can justify its trust in the restored environment with evidence, validation, and disciplined risk management. That is the difference between bringing systems back and truly recovering.

Critical Thinking Question: What evidence should an organization require before it treats a restored system or service as trustworthy again, rather than merely available?

---

[63] Michael Bartock, Jeffrey Cichonski, Murugiah Souppaya, Matthew Smith, Gregory Witte, and Karen Scarfone, Guide for Cybersecurity Event Recovery, NIST SP 800-184 (Gaithersburg, MD: National Institute of Standards and Technology, December 2016).

Part IV. Critical Infrastructure Defense

Changes in critical infrastructure alter cyber defense by changing what failure means. In these environments, disruption may affect public services, essential operations, regional stability, and systems that extend far beyond the organization experiencing the initial event. This section examines why critical infrastructure environments require a wider field of view, how IT and OT differ in mission and response implications, why ICS and OT incidents demand specialized handling, and how public-sector and essential-service constraints shape what is possible during a crisis. The purpose of this part is to move the reader beyond general cyber response and into the operational reality of defending systems whose consequences are broader, faster-moving, and often more publicly visible.

Chapter 14: Why Critical Infrastructure Changes Everything

Critical infrastructure changes the meaning of cyber defense because it changes the meaning of consequence. In ordinary enterprise environments, a serious cyber incident may still produce major financial loss, operational disruption, regulatory problems, and reputational harm. Those effects are real and can be severe. In critical infrastructure environments, however, the effects of disruption often extend beyond the organization itself into essential services, public confidence, supply continuity, emergency readiness, and regional stability. That broader consequence environment is what makes critical infrastructure defense a different professional lane rather than simply a more serious version of ordinary enterprise security.

CISA identifies 16 critical infrastructure sectors that together form a complex and interconnected ecosystem.[64] For purposes of this book, it is useful to understand them in practical terms rather than as a memorization exercise. The Chemical Sector includes the production, storage, and movement of industrial chemicals. The Commercial Facilities Sector includes venues and properties where people gather, conduct business, or receive services. The Communications Sector supports voice, data, and connectivity. The Critical Manufacturing Sector includes industrial production that supports broader national and economic function. The Dams Sector covers dam infrastructure and related water-control systems. The Defense Industrial Base Sector supports the industrial and supply relationships behind defense capability. The Emergency Services Sector includes law enforcement, fire, emergency medical, and related response functions. The Energy Sector includes electric power, oil, and natural gas

---

[64] Cybersecurity and Infrastructure Security Agency, "Critical Infrastructure Security and Resilience," accessed May 20, 2026, https://www.cisa.gov/topics/critical-infrastructure-security-and-resilience.

systems. The Financial Services Sector supports banking, payments, markets, and financial transactions.

The Food and Agriculture Sector includes the systems that support production, processing, and distribution of food. The Government Services and Facilities Sector supports the facilities and services used by government. The Healthcare and Public Health Sector includes healthcare delivery and public health support systems. The Information Technology Sector includes the digital systems, services, and platforms that underlie modern operations. The Nuclear Reactors, Materials, and Waste Sector includes civilian nuclear facilities and related materials management. The Transportation Systems Sector covers the movement of people and goods across multiple transportation modes. The Water and Wastewater Systems Sector supports drinking water, wastewater treatment, and related public health functions.[65]

Readers who want sector-specific definitions, designated Sector Risk Management Agencies, and current official materials should go directly to CISA's critical infrastructure sector pages for deeper study.[66]

That overview matters because the phrase critical infrastructure is sometimes used too casually. It does not simply mean "important IT." It refers to systems and sectors whose disruption can significantly affect public health, public safety, economic continuity, government capability, or daily life at scale. That difference should immediately change how defenders think about cyber incidents. In these environments, the central question is not merely whether systems have been compromised. It is whether the disruption of those systems changes the conditions under

---

[65] Cybersecurity and Infrastructure Security Agency, "Sector Risk Management Agencies," accessed May 20, 2026, https://www.cisa.gov/topics/critical-infrastructure-security-and-resilience/critical-infrastructure-sectors/sector-risk-management-agencies.
[66] Cybersecurity and Infrastructure Security Agency, "Critical Infrastructure Sectors," accessed May 20, 2026, through CISA critical infrastructure sector pages.

which communities, institutions, and dependent sectors continue to function.

This is why public consequence changes the response model. Once the organization is defending a function tied to water, fuel, communications, finance, transportation, health, or emergency support, the decision space becomes wider. A cyber issue is no longer only an internal security event. It may now have implications for continuity of service, public communication, mutual aid, external coordination, regulatory attention, sector-specific support, and broader resilience planning. The incident may still begin inside one organization's network, but the consequence environment is no longer confined to that organization's boundaries.

Interdependency is one of the main reasons this happens. Critical infrastructure systems rarely stand alone in practice, even when they appear administratively separate. Water systems depend on power, communications, transportation, chemicals, and industrial support. Emergency services depend on communications, transportation, fuel, IT, and facilities. Healthcare depends on power, water, IT, supply chains, and transportation. Finance depends on communications, IT, power, and continuity of operations. Transportation relies on fuel, communications, IT, and emergency services. A disruption in one sector may therefore produce second-order effects in others without every affected organization experiencing the same initial cyber event.[67] This is one of the defining characteristics of infrastructure defense. A local failure may become a regional problem because of dependency, not because the adversary directly compromised every affected system in the same way.

---

[67] Cybersecurity and Infrastructure Security Agency, "Learn," Infrastructure Dependency Primer, accessed May 20, 2026, https://www.cisa.gov/topics/critical-infrastructure-security-and-resilience/resilience-services/infrastructure-dependency-primer/learn.

CISA's infrastructure dependency work is useful here because it emphasizes that infrastructure sectors are linked by relationships that can transmit consequence across systems and organizations. For defenders, this means that cyber analysis must move beyond the compromised host or the immediate technical symptom. A useful question is not only, "What has been affected?" A more important question is often, "What depends on what has been affected, and how quickly will that dependence become operationally significant?" That is the point at which infrastructure cyber defense becomes much more than technical troubleshooting.

Time pressure also behaves differently in infrastructure environments. In many enterprise incidents, time matters because adversaries may expand access, steal more data, or deepen persistence while the organization delays. That remains true in critical infrastructure. But there is an additional clock. Services may not tolerate long interruptions. Backup modes may be sustainable for only limited periods. Operational tolerances may be narrow. Public expectations may be less forgiving. External events such as severe weather, supply disruption, emergency conditions, or regional instability may compress the decision timeline even further. The result is that defenders in infrastructure environments often operate under more than one clock at the same time: the attacker's clock, the organization's clock, and the service continuity clock.

This is one reason why infrastructure defense cannot be treated as a simple extension of ordinary enterprise practice. The systems involved are often older, more operationally constrained, more dependent on external support, and more tightly linked to real-world function.[68] Many include mixed IT and OT realities, legacy technologies, limited

---

[68] Kevin Stouffer et al., *Guide to Operational Technology (OT) Security*, NIST SP 800-82 Rev. 3 (Gaithersburg, MD: National Institute of Standards and Technology, September 2023).

maintenance windows, vendor-managed components, and uneven modernization. These are not signs of negligence. They are characteristics of environments built over time to support essential functions. A responder who enters this space expecting a clean, modern, and centrally governed environment will misread both the risks and the limits of available action.

NIST's OT security guidance reinforces this distinction by emphasizing that operational environments often have unique safety, performance, reliability, and availability requirements that differ from traditional IT priorities. That matters at the chapter level because it helps explain why infrastructure incidents resist simple cyber answers. A recommendation that is clearly sound in enterprise IT may be operationally dangerous in a process-dependent environment. A delay that seems tolerable from a technical perspective may become unacceptable from a service perspective. A narrow cyber fix may leave a broader dependency risk untouched. This is why infrastructure defense requires a wider field of view from the beginning.

That wider field of view is one of the most important professional adjustments in this book. The defender is no longer responsible only for identifying malicious activity and recommending technical remediation. The defender must also understand consequence, continuity, dependency, and the practical operating environment around the system at risk. This does not mean the responder becomes an engineer, regulator, executive, or emergency manager all at once. It means the responder must learn how to think beyond the immediate technical boundary and understand why the system matters in the first place.

For students and transitioning professionals, Chapter 14 is where the concept of critical infrastructure becomes more concrete. The importance of this field does not rest solely on abstract national security language. It rests on the fact that communities rely on these sectors every day, often without seeing the systems, personnel, or dependencies that make them function. Cyber incidents in these environments differ because the effects of failure can move quickly from technical abnormality to operational disruption, public consequences, and institutional stress. That broader effect makes the defender's job more demanding and consequential.

The practical lesson is straightforward. Critical infrastructure changes everything because it changes what failure means. It changes who is affected, how quickly consequences can spread, what forms of coordination become necessary, and what standards of judgment responders must apply before acting. If the defender approaches these environments as though they are simply high-stakes enterprise networks, the defender will underestimate the mission. If the defender understands them as interconnected systems of public and operational consequence, the rest of the incident response lifecycle becomes easier to interpret correctly.

Critical Thinking Question: Why can a cyber problem that appears local at first become a much broader crisis in a critical infrastructure environment, even when the initial technical symptoms seem limited?

Chapter 15: IT and OT Are Not the Same Fight

One of the fastest ways to make a critical infrastructure incident worse is to assume that enterprise IT logic transfers cleanly into operational technology. The assumption is understandable. Many defenders are trained first in environments where the primary mission is to protect information systems, user access, data, and business workflows. In those environments, common cybersecurity actions such as aggressive scanning, patching, forced reboots, account lockouts, rapid segmentation changes, and endpoint tooling are often treated as normal parts of defensive response. Those actions can be appropriate in many IT settings. The problem is that critical infrastructure often includes systems whose primary purpose is not information handling, but the monitoring and control of physical processes.[69]

That difference in purpose changes the entire risk equation. Information technology generally supports communication, storage, transactions, applications, identity, and administrative functions. Operational technology supports process execution, timing-sensitive control, physical state awareness, and, in many environments, the continuity of essential services. A disruption in one domain may be inconvenient or expensive. A disruption in the other may alter a treatment process, reduce process visibility, interrupt a control loop, impair operational confidence, or create safety and continuity problems that extend well beyond cybersecurity. This is why IT and OT are not simply different technical stacks. They are different mission environments.

NIST's guidance on operational technology makes this distinction explicit by emphasizing that OT systems interact with the physical environment and are governed by unique performance, reliability, and

---

[69] Kevin Stouffer et al., Guide to Operational Technology (OT) Security, NIST SP 800-82 Rev. 3 (Gaithersburg, MD: National Institute of Standards and Technology, September 2023).

safety requirements. That point has practical consequences for defenders. A cyber action that is low-risk in enterprise IT may be high-risk in OT if it affects timing, stability, interoperability, or availability in ways the physical process cannot tolerate. In these environments, the defender must think beyond whether an action is technically possible or technically justified. The defender must also ask whether the process can safely absorb the action.

This is why the traditional shorthand that "IT prioritizes confidentiality and OT prioritizes availability" is not wrong, but it is not sufficient. It oversimplifies a more important reality. OT-sensitive environments often require defenders to think in terms of process stability, deterministic behavior, operator visibility, fail-safe conditions, and safe continuity in addition to familiar cybersecurity concerns. Availability matters, but it matters in a particular way. It is not merely about uptime. It is about whether a process can continue to operate safely, predictably, and under appropriate control while the organization manages cyber risk. That makes some routine IT instincts less reliable in OT-affected environments.

A simple example illustrates the problem. In enterprise IT, forcing a restart on a suspicious system may be a reasonable first move if it cuts off a process, clears memory-resident activity, or supports recovery sequencing. In an OT-dependent environment, the same action may be destabilizing if the system supports process visibility, alarm functions, timing-sensitive control, or communication with devices that do not recover gracefully. Similarly, broad credential resets may strengthen security in principle, but if they are applied without understanding which accounts support operational maintenance, process continuity, or interdependent systems, they can create operational confusion at exactly the wrong moment. The issue is not that IT habits are useless. The issue is that they require more consequence-aware discipline before being applied.

Visibility is also different. Many enterprise defenders assume that if something matters, they can install an agent, collect telemetry, run scans, or increase logging quickly enough to create a workable picture. That assumption is often unreliable in OT environments. Devices may be fragile. Protocols may be specialized. Vendor support may be restrictive. Legacy systems may not tolerate modern security tooling well. Some environments may have limited native logging, limited centralized collection, or very narrow maintenance windows in which changes are safe.[70] That does not mean defenders should give up on visibility. It means they should approach visibility as a constrained engineering and operational problem, not as a default technical entitlement.

CISA's industrial control systems resources reinforce this practical point by treating OT and ICS cybersecurity as a distinct domain requiring specialized collaboration, risk reduction, and operational awareness.[71] Defenders should read that as a warning against overconfidence. In many infrastructure environments, the cyber team will not hold the full picture alone. Operators, engineers, maintenance personnel, vendors, and control-system specialists may each understand parts of the process that the cybersecurity team cannot infer from network diagrams or asset lists. If those perspectives are not brought together, the organization may end up making cyber decisions that are technically sound and operationally damaging.

This is why critical infrastructure defense depends so heavily on collaboration across disciplines. Cyber teams understand attacker behavior, access risk, monitoring, identity, and containment logic.

---

70 Cybersecurity and Infrastructure Security Agency, ICS Recommended Practices, accessed May 20, 2026, https://www.cisa.gov/resources-tools/resources/ics-recommended-practices; Cybersecurity and Infrastructure Security Agency, Industrial Control Systems, accessed May 20, 2026, https://www.cisa.gov/topics/industrial-control-systems.

71 Cybersecurity and Infrastructure Security Agency, Industrial Control Systems, accessed May 20, 2026, https://www.cisa.gov/topics/industrial-control-systems.

Engineers and operators understand process behavior, system tolerance, control priorities, and the physical meaning of disruption. Neither view is sufficient by itself. A responder entering this environment must therefore become comfortable working inside a team where not everyone speaks the same technical language and where “secure” does not always mean “operationally viable” without additional coordination.

The rise of zero trust concepts in OT underscores this same lesson. CISA’s 2026 guidance on adapting zero trust principles to operational technology does not suggest that OT environments can simply import enterprise zero trust patterns unchanged.[72] Instead, it recognizes that legacy architecture, operational constraints, and process-critical dependencies require adaptation. That is the right mindset for this chapter as well. The goal is not to force OT into an IT defensive mold. The goal is to apply strong security principles in a way that respects the actual operating conditions of the environment.

This is another important shift for readers moving from cyber warfighter to cyber defender roles. In some highly centralized or mission-driven contexts, there can be an instinct to prioritize decisive technical action above all else. In civilian critical infrastructure environments, that instinct must be tempered by a different responsibility. The defender is not only responsible for reducing cyber risk. The defender is also responsible for ensuring that the means of reducing that risk do not create a worse operational outcome. That is not hesitation. It is mature defensive judgment.

The practical question changes accordingly. In ordinary IT response, defenders often ask, “How fast can we isolate, clean, and restore?” In OT-sensitive environments, the better question is often, “How

---

[72] Cybersecurity and Infrastructure Security Agency, Adapting Zero Trust Principles to Operational Technology, April 29, 2026, https://www.cisa.gov/resources-tools/resources/adapting-zero-trust-principles-operational-technology.

do we reduce attacker opportunity without destabilizing the process the organization still depends on?" Those questions overlap, but they are not identical. The second question forces the responder to think in terms of consequence, sequencing, coordination, and safe change. It also forces the defender to respect that some parts of the environment cannot be touched as casually as others.

This is why IT and OT are not the same fight. They are connected, sometimes tightly, but they are governed by different tolerances and different definitions of harm. A defender who approaches OT with IT reflexes alone is likely to overestimate what can be changed quickly, underestimate the operational meaning of disruption, and misunderstand why engineering and operations may resist technically aggressive actions. A defender who understands the difference will still act decisively when necessary, but with a broader and more responsible view of what the action is likely to do.

In critical infrastructure, the safest technical move is not always the safest operational move. Understanding that distinction is one of the essential professional transitions in this field.

Critical Thinking Question: What kinds of cyber response actions become more dangerous when the affected system supports a physical process, and how should a defender decide when operational consequences outweigh the appeal of a fast technical fix?

Chapter 16: Incident Response in ICS and OT Environments

Incident response in industrial control systems and operational technology environments is not simply enterprise incident response applied to more fragile hardware. It is a different operational problem. The systems involved interact with the physical world, support timing-sensitive processes, and often operate under reliability and safety expectations that do not tolerate the same kinds of disruption commonly accepted in ordinary IT response. NIST states that OT security must address these unique performance, reliability, and safety requirements because OT systems monitor or control devices, processes, and events in the physical environment.

That difference changes the meaning of the response from the moment the organization suspects that ICS- or OT-relevant systems may be involved. In enterprise IT, a responder may focus primarily on protecting information systems, limiting adversary access, preserving business functionality, and restoring services. In ICS and OT environments, those goals still matter, but they exist alongside a different set of questions. Can the process continue to operate safely? What functions cannot be interrupted without creating a worse outcome? What systems support visibility into the physical process, even if they are not themselves the source of compromise? What dependencies are poorly understood but operationally essential? These are not secondary questions. In many cases, they are the most important questions in the room.

This is why safety and stability must shape every major response decision. A technically aggressive action may be entirely appropriate in an enterprise environment and entirely inappropriate in a process-dependent environment. A rapid reboot, broad network isolation, aggressive scanning campaign, or blanket credential change may reduce cyber risk in one sense while increasing operational risk in another. NIST's OT security guidance and CISA's ICS resources both reinforce the same practical lesson: defenders in these environments must reduce cyber risk without undermining the reliability and safety of the industrial process itself.

This requirement immediately expands the response team. Cyber teams cannot work alone once the incident touches systems that influence process monitoring, control, or continuity. Operators, engineers, maintenance leaders, plant or facility managers, and sometimes vendors must be involved because they understand process tolerance, fallback conditions, sequencing constraints, and the operational meaning of interruption. The cyber team provides attacker awareness, access path analysis, monitoring logic, and containment discipline. Engineering and operations bring process understanding. Neither perspective is sufficient by itself. If the organization allows one to dominate without the other, it increases the chance of making a technically correct decision that is operationally harmful, or an operationally cautious decision that leaves the attacker too much room to persist.

This is one reason response in ICS and OT environments is often staged rather than absolute. Full and immediate isolation may sometimes be the right choice, but many real situations demand more selective action. The organization may need to restrict remote access while preserving local operations. It may need to move into manual or degraded operation before it can safely sever a digital dependency. It may need to increase monitoring around one zone while preparing engineering staff for a later containment step in another. It may need to sequence control changes carefully so that the process remains visible and stable while trust is rebuilt. CISA's recommended practices for ICS and broader ICS guidance support this kind of structured and risk-aware approach rather than a reflexive "disconnect first, sort it out later" mindset.

Visibility also becomes more constrained in these environments, and responders need to accept that limitation without surrendering professional responsibility. In many OT environments, defenders cannot simply deploy agents, increase logging everywhere, or run the same kinds of active discovery tools they would use in enterprise IT. Devices may be legacy systems. Protocols may be brittle. Vendors may limit what can be changed. Maintenance windows may be narrow. Logging may be incomplete. In some environments, the organization may have better visibility into the business network than into the process network that matters most during the incident. NIST's OT guidance and CISA's ICS recommended practices both make clear that defenders should account for these realities rather than assume they can force traditional IT visibility into systems that were never designed for it.

This constraint makes manual operations, degraded modes, and fallback planning much more important than many IT-focused responders initially expect. In some ICS and OT incidents, the organization preserves safety and continuity not by keeping every digital pathway intact, but by shifting temporarily into more controlled, less automated modes of operation. That shift may slow the process, increase labor demands, reduce efficiency, and require tighter coordination, but it can also buy the organization the time needed to restrict untrusted pathways and validate what can be safely restored. CISA's ICS best-practices materials explicitly call for maintaining and testing incident response plans in industrial environments, which reflects the broader point that fallback is not improvisation if it has been prepared and exercised in advance.

Another important difference is that incident success in these environments cannot be measured only by the absence of visible malicious activity. In enterprise response, teams sometimes speak loosely about getting the environment "clean" once obvious indicators have been removed and services are coming back. In ICS and OT environments, that language is often too imprecise. The better question is whether the process is stable, observable, and trustworthy enough to support continued operation. If operators do not trust the readings, if engineers do not trust the support pathways, if responders cannot explain why the environment is now safer than before, or if the organization cannot sustain the compensating controls required to keep operating, then the response is not truly complete. It may be quieter, but it is not yet reliable.

This chapter also reinforces the professional shift from cyber warfighter to cyber defender. In civilian critical infrastructure environments, the mission is not to outmaneuver the attacker in an unconstrained technical contest. The mission is to protect essential operations lawfully, carefully, and effectively. That means defending the process, not just the network. It means accepting that some of the most mature response decisions are the ones that preserve safe operations while reducing adversary opportunity in stages. It means understanding that industrial response is not slower because the defenders are weaker. It is often more deliberate because the consequences of a wrong move are broader.

CISA's current ICS resources, ICS training offerings, and recent OT-focused zero trust guidance all point in the same direction: ICS and OT environments require specialized defensive thinking, stronger collaboration across disciplines, and response methods that align with operational realities rather than abstract cyber ideals. The practical lesson is straightforward. If the responder does not understand what the system does, what the process can tolerate, and what the organization must keep stable while it responds, then the responder is not yet prepared to lead or materially shape incident response in that environment.

The right response in ICS and OT environments is therefore not the one that looks most decisive from a narrow cybersecurity perspective. It is the one that reduces attacker opportunity while preserving the safety, stability, and continuity of the process the organization still depends on.

Critical Thinking Question: How should a responder decide when a cyber containment action in an ICS or OT environment reduces overall risk, and when it instead creates a new and more dangerous operational problem?

Chapter 17: Public Sector and Essential Service Constraints

Critical infrastructure defense does not occur in an ideal operating environment. It occurs inside institutions that are often constrained by staffing limitations, vendor dependencies, procurement timelines, overlapping authorities, public accountability, and the practical difficulty of coordinating among organizations that share risk without sharing command. These realities are not side issues. They shape what can be done, how quickly it can be done, and who must be involved before the organization can move from technical understanding to operational action.

One of the most persistent errors in cybersecurity writing is the assumption that good decisions naturally produce timely execution. In reality, the organization may know what it needs to do and still be unable to do it at the speed the incident demands. A critical owner may be unavailable. A vendor may need to approve or execute part of the change. A dependency may sit inside another team's area of responsibility. A public-sector entity may need to work through formal channels, documented approvals, or interagency coordination processes before a technically straightforward recommendation can become an operational reality. The issue is not always indecision. Sometimes it is institutional friction.

Staffing is one of the most important parts of that friction. Cybersecurity workforce shortages are not abstract labor-market problems for public-sector and essential-service organizations. They directly affect incident readiness, continuity of expertise, after-hours coverage, and the ability to sustain response over time. GAO reported in 2025 that building and maintaining a cybersecurity workforce is vital to protecting government operations, but that a shortage of skilled workers has made this challenging across departments it reviewed.[73] In practice, that kind of staffing pressure means that some organizations rely heavily on a small number of key people, operate with thin bench depth, or struggle to maintain the same quality of response during nights, weekends, holidays, or prolonged incidents. A process that appears viable on paper can become fragile very quickly if too much of it depends on too few people.

This matters even more in essential-service environments because incident response rarely occurs in isolation from normal operations. The same people who understand the environment may also be supporting routine service continuity, maintenance coordination, vendor communications, compliance obligations, or executive updates. During a major incident, that means the organization is not only trying to investigate and respond. It is also trying to continue operating. When staffing is thin, those missions compete with each other. The result is that even highly competent teams may find themselves forced to triage not only the incident, but their own attention.

The workforce issue also has a strategic dimension. GAO's 2025 High-Risk Series noted that federal agencies still face challenges in hiring, training, and retaining cybersecurity and privacy workforces, including challenges that affect critical infrastructure entities.[74]That observation matters beyond the federal context. Public-sector and essential-service organizations often compete for cybersecurity talent against private-sector employers that can offer faster hiring, different compensation structures, or less institutional complexity. The practical effect is that some infrastructure defenders must protect high-consequence systems without the staffing

---

[73] U.S. Government Accountability Office, Cybersecurity Workforce: Departments Need to Fully Implement Key Practices, GAO-25-106795 (Washington, DC: January 16, 2025).
[74] U.S. Government Accountability Office, High-Risk Series (Washington, DC: February 25, 2025), discussion of federal cybersecurity workforce management challenges affecting critical infrastructure entities.

depth that the mission would ideally require. Mature leadership does not ignore that constraint. It plans around it.

Coordination is the second major source of friction. Critical infrastructure incidents often involve entities that must work together even though they do not report to one another, do not share the same tools, and do not operate under the same priorities. FEMA's National Incident Management System explicitly describes the need for all levels of government, non-governmental organizations, and the private sector to work together during incidents.[75] That principle applies strongly in cyber incidents affecting essential services. Utilities, local governments, public safety agencies, private vendors, managed service providers, state agencies, federal partners, and critical service operators may all become part of the same consequence picture. Shared consequence does not create shared authority, and that is one reason response can slow at exactly the point where urgency is increasing.

This is where many responders become frustrated for the wrong reason. They assume the problem is that others do not understand the seriousness of the technical issue. Sometimes that is true. Often, however, the deeper problem is that different participants are carrying different responsibilities inside the same incident. The cyber team may be focused on access paths, scope, containment options, and preservation of evidence. Operations may be focused on maintaining service continuity and safe process behavior. Legal may be focused on reporting obligations, contractual exposure, and disciplined language. Executive leadership may be focused on prioritization, public consequence, and the political or institutional implications of disruption. None of these perspectives cancels

---

[75] Federal Emergency Management Agency, National Incident Management System (3rd ed., Washington, DC: U.S. Department of Homeland Security, 2017); see also Federal Emergency Management Agency, "National Incident Management System," updated July 28, 2025, https://www.fema.gov/emergency-managers/nims.

the others. The difficulty lies in getting them aligned quickly enough to support action.

Public accountability intensifies this challenge. Critical infrastructure incidents in public-sector or essential-service contexts often unfold under greater visibility than ordinary enterprise incidents. The organization may have to consider public trust, elected oversight, media attention, records obligations, regulator expectations, and sector-specific scrutiny while the incident is still developing. CISA's critical infrastructure mission materials emphasize the role of supporting state, local, tribal, territorial, and industry partners in maintaining the functions the public depends on.[76] That broader mission means the organization cannot treat response as a purely internal technical matter. The consequences of communication failure, poorly framed decisions, or visible service disruption may be larger and more durable than the immediate technical event.

This does not mean public-sector and essential-service organizations are inherently weaker than private-sector organizations. It means they often operate inside a more visibly constrained decision environment. When public consequence is attached to the service, the organization may need to weigh transparency, defensibility, and continuity more explicitly in parallel with cyber action. A recommendation that would be simple inside a private internal network may require more careful evaluation when the affected function is tied to public health, water reliability, emergency support, transportation continuity, or another visible service. Responders who do not understand this institutional reality will often mistake disciplined coordination for hesitation.

Vendor and external support dependencies add another layer. Many infrastructure operators and public entities rely on external

---

[76] Cybersecurity and Infrastructure Security Agency, "Critical Infrastructure Security and Resilience," accessed May 20, 2026, https://www.cisa.gov/topics/critical-infrastructure-security-and-resilience.

technology providers, managed services, industrial vendors, or support contracts for essential technical functions. Those relationships can be necessary and well-managed while still limiting the tempo of response. The organization may know what it needs to change but lack immediate authority, access, expertise, or contractual clarity to execute that change without outside participation. In some cases, vendor support is part of resilience. In other cases, it becomes part of the friction that slows adaptation under pressure. The key issue is not whether vendors are involved. It is whether the organization has prepared for the reality that third-party involvement can shape the speed and flexibility of response.

CISA's incident response materials reinforce that responding to significant cyber incidents involves contributions from the private sector and from local, state, and federal entities rather than a single isolated actor.[77] That observation is useful because it captures the deeper truth of Chapter 17: the responder is working inside a system, not merely inside a network. The more consequential the service, the more likely it is that effective response will require coordination across boundaries the cyber team does not control directly.

For professionals moving from a more centralized cyber role into critical infrastructure defense, this chapter marks another important shift. The defender must stop assuming that correct technical analysis is enough to move the environment. In these settings, strong technical work remains essential, but it must be paired with institutional awareness. The responder must understand staffing realities, ownership friction, dependency chains, communication structure, and the fact that some delays arise not from ignorance but from the way the environment is built

---

[77] Cybersecurity and Infrastructure Security Agency, "Incident Response," accessed May 20, 2026, https://www.cisa.gov/topics/cyber-threats-and-response/incident-response.

to govern risk. That does not make the work less serious. It makes it more realistic.

The practical question is not whether these constraints are frustrating. They are. The more important question is whether the organization understands them well enough to prepare around them. Mature defenders do not waste time pretending the institution is frictionless. They learn where staffing is thin, where coordination slows, where vendors must be engaged, where authority is fragmented, and where public consequence changes the decision process. That awareness does not solve every problem. It does, however, make better preparation and better response possible.

Essential-service incident response is therefore shaped as much by institutional reality as by technical capability. Responders who ignore that reality will repeatedly be surprised by delays, ownership gaps, and decision bottlenecks. Responders who understand it can begin to work on the actual problem rather than the idealized one.

Critical Thinking Question: How can a defender improve incident response performance in a public-sector or essential-service environment without pretending that staffing limits, coordination friction, vendor dependence, and public accountability do not exist?

Part V. Leadership and Continuous Improvement

No incident response capability is complete without leadership, communication, and long-term learning. Technical skills alone cannot keep an organization aligned during a crisis, nor can recovery alone guarantee that the next incident will begin in a stronger environment. This section focuses on leading during the incident, communicating across technical and nontechnical audiences, conducting after-action reviews that produce real change, and building long-term resilience in critical infrastructure defense. The purpose of this part is to close the book by reinforcing that enduring capability is built not only through response but through disciplined leadership, institutional learning, and the willingness to turn disruption into lasting improvement.

Chapter 18: Leading During the Incident

Leadership during a cyber incident is often misunderstood as visibility, authority, or decisiveness alone. In practice, leadership is the disciplined creation of clarity in conditions where clarity is naturally scarce. The incident may be technically complex, operationally disruptive, politically sensitive, or publicly consequential, but the organization still has to function. Teams still need priorities. Decision-makers still need a common understanding of what matters now. Stakeholders still need information framed in a way they can act on. In that environment, leadership is not a ceremonial layer added on top of response. It is one of the mechanisms that keeps the response from fragmenting under pressure.[78]

This is why leadership begins before certainty arrives. One of the most common mistakes in crisis settings is to assume that leaders should wait until the technical picture is stable before imposing structure. Real incidents do not allow that luxury. A leadership vacuum in the early phase of an incident rarely stays empty. It is usually filled by competing assumptions, unmanaged urgency, duplicated requests, and functionally isolated decisions made without enough shared context. NIST's current incident response guidance explicitly connects incident response to broader cybersecurity risk management and to the organizational activities that make response more effective.[79] That broader view is important because it recognizes that technical work alone does not keep the organization coherent. Leadership does.

---

[78] Federal Emergency Management Agency, "National Incident Management System," updated July 28, 2025, https://www.fema.gov/emergency-managers/nims.
[79] Alexander Nelson, Sanjay Rekhi, Murugiah Souppaya, and Karen Scarfone, Incident Response Recommendations and Considerations for Cybersecurity Risk Management: A CSF 2.0 Community Profile, NIST SP 800-61 Rev. 3 (Gaithersburg, MD: National Institute of Standards and Technology, April 2025), https://doi.org/10.6028/NIST.SP.800-61r3.

The first practical product of leadership is clarity. Clarity does not mean certainty. It means defining what is known, what is not yet known, what decisions are pending, which risks appear most urgent, and who owns the next set of actions. This is especially important in critical infrastructure environments, where technical teams, operators, legal counsel, executives, vendors, and continuity personnel may all be looking at the same incident through different responsibility lenses. If leadership does not reduce that ambiguity into a usable decision structure, the organization may still be busy, but it will not be coordinated.

This is where priority setting becomes essential. Incidents produce more urgent demands than most organizations can satisfy simultaneously. Technical teams want time to validate indicators and improve scope. The operations team wants to understand what functions can remain available. Executives want to know which decisions must be made now and which can wait. Legal and communications teams want disciplined language and defensible claims. External partners may need timely notification or coordination. FEMA's National Incident Management System emphasizes common operating picture and management-by-objectives because response breaks down quickly when organizations cannot establish shared priorities and a common understanding of the incident.[80] That principle applies directly to cyber incidents. If everything is treated as equally urgent, then nothing is actually prioritized.

Leadership also matters because it protects the response from unnecessary noise. In many serious incidents, technical teams become overloaded not only by the incident itself but by the organization's reaction to it. Repeated requests for updates, conflicting direction, poorly timed

---

80 Federal Emergency Management Agency, "National Incident Management System," updated July 28, 2025, https://www.fema.gov/emergency-managers/nims; Federal Emergency Management Agency, National Incident Management System (3rd ed., Washington, DC: U.S. Department of Homeland Security, 2017).

demands for certainty, and direct stakeholder pull on the investigation can all degrade response quality. A mature leader does not eliminate the need for information. A mature leader creates rhythm. Briefings happen at known intervals. Questions are routed through defined channels. Decisions are identified explicitly rather than hidden inside general status requests. The technical team is protected from avoidable disruption without isolating it from the organization it serves.

This is one reason leadership should not be confused with interference. Some leaders remain highly visible during an incident but contribute little to the quality of the response because they substitute constant presence for useful direction. They attend every technical discussion, request frequent informal updates, and inject urgency without narrowing the decision space. That behavior can look engaged while making the organization less effective. Strong leadership is different. It stays connected to the incident while preserving room for technical work, operational coordination, and disciplined communication. CISA's incident response planning materials and tabletop exercise resources both reflect this broader point by emphasizing defined roles, responsibilities, and structured participation rather than ad hoc involvement by whoever happens to have authority.[81]

Critical infrastructure environments make this leadership problem more demanding because the response must hold together more than one reality at once. A cyber team may be concerned with containment, persistence, and scope. Operations may be concerned with continuity and process stability. Executives may be concerned with public consequence,

---

[81] Cybersecurity and Infrastructure Security Agency, Incident Response Plan (IRP) Basics, January 31, 2024, https://www.cisa.gov/resources-tools/resources/incident-response-plan-irp-basics; Cybersecurity and Infrastructure Security Agency, Federal Government Cybersecurity Incident and Vulnerability Response Playbooks, updated 2024, https://www.cisa.gov/resources-tools/resources/federal-government-cybersecurity-incident-and-vulnerability-response-playbooks.

mission impact, and institutional exposure. External agencies or partners may be concerned with sector coordination or mutual support. None of these views is irrelevant. The leader's job is not to erase the distinctions. It is to keep them aligned well enough that the organization continues to move in one direction rather than splitting into functional silos with competing priorities.

This is another point where the transition from cyber warfighter to cyber defender becomes visible. In more centralized or mission-driven environments, there can be an expectation that once the threat is understood technically, the appropriate action should follow directly. In civilian critical infrastructure settings, the relationship between technical truth and organizational action is more complicated. Leadership must integrate technical risk, operational consequence, legal boundaries, continuity pressure, public accountability, and institutional constraint in real time. That does not mean leadership becomes vague or passive. It means leadership becomes more explicitly responsible for balancing multiple legitimate forms of risk at once.

Calm is part of this work, but it should not be misunderstood. Calm does not mean emotional distance or lack of urgency. It means controlled behavior that preserves capacity inside the organization. A leader who appears panicked, erratic, or overly reactive increases cognitive load across the entire response. A leader who communicates clearly, holds a disciplined cadence, and resists the demand for premature certainty makes it easier for others to perform well. In that sense, calm is not a personality trait. It is an operational asset.

Exercises and structured practice reinforce this point. CISA's tabletop exercise packages exist in part because organizations perform better when roles, decisions, and communication patterns have been tested before a real event forces them into motion.[82] That is relevant to leadership because effective crisis leadership is rarely improvised successfully. It depends on habits of framing, prioritization, briefing, and role discipline that have been developed in advance. Leaders who wait until a live incident to discover how their organization communicates, decides, and escalates are operating too late.

This is why leadership during the incident should be judged by function rather than by posture. Did the leader create clarity? Did the leader establish priorities? Did the leader help the organization understand which decisions mattered most? Did the leader protect the response from unnecessary noise? Did the leader force useful discipline into communications and cadence? Did the leader keep technical, operational, and institutional realities aligned well enough that the organization could act? Those are the questions that matter. Titles, presence, and verbal confidence are secondary if they do not improve those outcomes.

In critical infrastructure defense, leadership is not the act of sounding certain. It is the act of helping the organization function under uncertainty without losing discipline, focus, or direction. That is what makes leadership operationally real during an incident rather than merely managerial in appearance.

Critical Thinking Question: What specific leadership behaviors improve the quality of incident response under pressure, and which common leadership habits create the appearance of control while actually making the organization less effective?

---

[82] Cybersecurity and Infrastructure Security Agency, “CISA Tabletop Exercise Packages,” accessed May 20, 2026, https://www.cisa.gov/resources-tools/services/cisa-tabletop-exercise-packages.

Chapter 19: Communication Across Audiences

Communication during a cyber incident is often treated as a supporting activity rather than as part of the response itself. That is a mistake. In practice, communication shapes the incident because it shapes who understands the problem, who acts on it, how quickly decisions are made, and whether the organization remains aligned under pressure. A technically capable team can still underperform if the right people do not receive the right information in a form they can use. Conversely, a disciplined communication structure can improve the quality of response even when the technical picture remains incomplete.[83]

This is especially important in critical infrastructure environments because incidents do not unfold only inside technical teams. They unfold across operators, executives, legal advisors, public information staff, continuity planners, regulators, vendors, and external partners who each need some version of the same truth but not the same level of technical detail. The challenge is not merely to "keep everyone informed." The challenge is to preserve a common operating picture while tailoring information to the decision or action each audience must take.

A technical team may need indicators, logs, account activity, timing, system relationships, and investigative uncertainty. An executive may need to understand service impact, risk trajectory, decision deadlines, and what options exist. Legal or compliance staff may need to know what has been confirmed, what language should be avoided, and what obligations may be triggered. Operations leaders may need to know what systems, processes, or dependencies could be affected if the organization acts or delays. Public information personnel may need enough clarity to support disciplined external communication without overstating what is known.

---

[83] Cybersecurity and Infrastructure Security Agency, Incident Response Plan (IRP) Basics (Washington, DC: U.S. Department of Homeland Security, January 31, 2024), https://www.cisa.gov/resources-tools/resources/incident-response-plan-irp-basics.

These are different needs, but they should not produce different versions of the incident. They should produce different expressions of the same disciplined understanding.

This is one reason a common operating picture matters so much. FEMA's National Incident Management System emphasizes the importance of a common operating picture because organizations cannot coordinate effectively if they are making decisions inside conflicting narratives of the same event.[84] In cyber response, a common operating picture does not mean that every participant receives the same technical briefing. It means that the facts, uncertainties, priorities, and current decision points are aligned across the people who must act. Without that alignment, communication volume can increase while decision quality declines.

The most useful communication structure during an incident is often the simplest one. Responders should be able to explain what is known, what is unknown, what is being done, and what matters next. That basic structure helps technical teams avoid overstating early conclusions, helps leaders avoid demanding false certainty, and helps the organization distinguish between confirmed facts and active investigation. It also reduces one of the most damaging communication habits in cyber incidents: allowing speculation, assumptions, and incomplete observations to circulate as though they were equivalent to confirmed information.

This is where many organizations lose clarity. A technical detail is shared out of context. A low-confidence observation is repeated without its uncertainty attached. A tentative hypothesis becomes the basis for external messaging. An executive asks for a simple answer to a complex

---

[84] Federal Emergency Management Agency, National Incident Management System (3rd ed., Washington, DC: U.S. Department of Homeland Security, 2017); see also Federal Emergency Management Agency, "National Incident Management System," updated July 28, 2025, https://www.fema.gov/emergency-managers/nims.

question, and the response team gives one for the sake of speed rather than accuracy. Each of these failures can create downstream problems. They can distort prioritization, force premature commitments, damage trust, or lead the organization into actions based on an unstable understanding of the incident.

For that reason, communication discipline is not only about speaking clearly. It is also about speaking carefully. NIST's current incident response guidance supports this by treating communications and coordination as core parts of incident response preparation and execution rather than as optional supporting functions.[85] The responder should know when to brief, what to brief, what level of confidence to attach to the information, and which audience needs which form of that information. A mature team understands that a technically correct statement can still be poorly communicated if it is delivered without consequence framing, without confidence boundaries, or without attention to the needs of the audience.

This is particularly important during compound crises. In essential-service environments, a cyber incident may develop alongside weather events, physical disruptions, public emergencies, political pressure, or media attention. As those pressures increase, the number of urgent questions around the incident also increases. Some of those questions will be operationally relevant. Others will be adjacent, emotionally urgent, or outside the responder's actual lane. A disciplined communication model helps the organization avoid being pulled into every surrounding issue at once. It creates a way to identify what belongs inside the cyber response decision space, what belongs in continuity or emergency management

---

[85] Alexander Nelson, Sanjay Rekhi, Murugiah Souppaya, and Karen Scarfone, Incident Response Recommendations and Considerations for Cybersecurity Risk Management: A CSF 2.0 Community Profile, NIST SP 800-61 Rev. 3 (Gaithersburg, MD: National Institute of Standards and Technology, April 2025), https://doi.org/10.6028/NIST.SP.800-61r3.

channels, and what must be routed to public affairs, legal, or executive leadership rather than answered casually by the technical team.

That last point deserves emphasis. Not every urgent question is the responder's question to answer. This is one of the most important communication habits in critical infrastructure defense. When the incident is unfolding inside a larger environment of public consequence, stakeholders may ask for predictions, certainty, validation of rumors, statements about external effects, or explanations about topics outside the technical team's authority. A disciplined responder should not improvise answers simply because the moment feels urgent. Instead, the responder should acknowledge what the team knows, explain what remains outside its confirmed knowledge, and ensure that issues are routed to the functions that actually own them.[86] That is not avoidance. It is role discipline.

Internal communication should also protect the technical response from unnecessary noise. During serious incidents, technical teams can be overwhelmed by requests for updates, repeated requests for the same information, and questions that are not actually tied to a decision. This is where leadership and communication structure intersect. If every stakeholder pulls directly on the technical team, the organization may believe it is improving awareness while actually degrading response quality. A stronger model channels those demands through a defined briefing rhythm, common status format, and known points of contact so that technical teams can keep working while leadership still receives the information it needs.

External communication requires even more discipline. Essential-service organizations may need to communicate with regulators, sector

[86] Cybersecurity and Infrastructure Security Agency, "Incident Response," accessed May 20, 2026, https://www.cisa.gov/topics/cyber-threats-and-response/incident-response.

partners, local or state authorities, customers, the public, or the media. In those situations, the organization should be especially careful to distinguish confirmed operational impact from unresolved technical assessment. FEMA's public information guidance under the NIMS framework emphasizes the need for coordinated public information mechanisms because fragmented messaging creates confusion and damages trust.[87] In cyber incidents, this means the organization should avoid making public statements that imply technical certainty it does not yet possess or service confidence it has not yet validated. It is better to communicate cautiously and accurately than to overstate stability and later retract it.

This does not mean communication should be vague. Vague communication creates its own problems. Stakeholders need useful information. What they usually need most is not a flood of technical detail, but clear answers to a practical set of questions: What is confirmed? What functions are affected? What is being done? What decisions are pending? What should this audience do now, if anything? What remains under investigation? Those questions keep communication actionable rather than performative.

For readers transitioning from more centralized cyber roles into critical infrastructure defense, this chapter reinforces another professional shift. Communication is not an administrative afterthought to technical work. It is one of the ways technical work becomes operationally meaningful. A responder who cannot translate technical risk into disciplined, audience-appropriate communication leaves the organization less able to act, even if the technical analysis itself is correct. In

---

[87] Federal Emergency Management Agency, "Public Information and Warning," National Incident Management System training and public information resources, accessed May 20, 2026; see also FEMA Emergency Management Institute public information materials.

distributed, high-consequence environments, that communication skill is not optional. It is part of the craft.

The practical lesson is simple. Good communication during an incident does two things at once. It helps the right people act, and it protects the response from losing focus. It preserves shared understanding without pretending certainty, and it maintains role boundaries without becoming evasive. In high-consequence cyber defense, that is not a soft skill layered on top of real work. It is part of the real work.

Critical Thinking Question: How can responders communicate clearly enough to support action during an incident without overstating what is known, creating false confidence, or taking ownership of questions that belong to other functions?

Chapter 20: Effective After-Action Review

Most organizations conduct some form of review after a serious incident. The problem is not that reviews fail to occur. The problem is that many of them fail to produce meaningful improvement. A timeline is assembled, key events are summarized, and a small set of observations is recorded, but the review stops short of asking the harder questions. Why did the organization interpret the situation the way it did? Which assumptions shaped the earliest decisions? What forms of friction slowed action? Which communication patterns helped or harmed response quality? What weaknesses were revealed in identity management, dependency awareness, staffing depth, continuity planning, or leadership structure? A review that cannot answer those questions may still produce documentation, but it does not necessarily produce readiness.

This matters because the purpose of an after-action review is not to create a cleaner story about the incident. Its purpose is to improve the organization's next performance. NIST's current incident response guidance places incident response inside broader cybersecurity risk management and stresses that organizations should improve the efficiency and effectiveness of detection, response, and recovery activities over time. FEMA's continuous improvement guidance makes a similar point from the exercise and real-world incident side: an after-action effort should identify strengths, areas for improvement, and recommended actions, and those actions should be monitored and implemented as part of improving preparedness. In other words, a review has value only if it changes future behavior, future structure, or future decisions.

One reason many reviews underperform is that they rely too heavily on chronology. Timelines are necessary. They help the organization reconstruct what happened, when it happened, and how decisions aligned with visible developments. But chronology is not the same thing as explanation. Knowing that a system was isolated at a certain time or that a briefing occurred at a certain hour does not explain why that choice was made, why it took that long, what information was missing, or whether the decision was reasonable under the conditions at the time. A strong after-action review therefore uses the timeline as a foundation, not as the end product. It asks what the timeline reveals about judgment, assumptions, role clarity, and organizational performance under stress.

The most useful review questions focus on decision points rather than on events alone. An organization should examine why weak indicators were or were not treated as significant early enough, why scope was framed narrowly or broadly, why containment choices were sequenced the way they were, why leadership communicated in the language it used, and why specific teams had or lacked the authority they needed. The point is not to second-guess every action with hindsight arrogance. The point is to understand how the organization thought during the incident and whether that way of thinking should change before the next one. That is a more demanding task than assembling facts, but it is also the task that produces real learning.

This is where many organizations encounter the difference between root cause and contributing factors. Serious incidents are rarely explained by one failure alone. A missed alert may matter, but so may fragmented ownership, weak identity controls, poor asset understanding, untested continuity assumptions, vendor dependence, or unclear escalation thresholds. If the review reduces the event to a single obvious mistake, it may create the false comfort of simplicity while leaving the larger pattern untouched. FEMA's improvement-planning doctrine and continuous improvement guidance both reinforce the idea that findings should connect to corrective actions and capability improvement rather than to superficial narrative closure. That principle applies equally to cyber incidents. The organization should identify what directly caused failure, what enabled it, what amplified it, and what made recovery harder than it should have been.

A serious after-action review also requires accountability without turning into blame theater. If the review becomes an effort to identify one person or one team to carry the burden of a complex event, the organization will learn very little. People will protect themselves, nuance will disappear, and structural weaknesses will be recast as individual failure. On the other hand, a review that avoids ownership altogether becomes equally useless. If no one is responsible for fixing stale contact paths, weak escalation practices, thin identity controls, poor dependency mapping, or broken briefing structure, then the review is just a better-organized memory. The right balance is disciplined accountability. The organization should identify where responsibility sat, whether that responsibility was realistic under the circumstances, and who now owns the corrective action required to improve performance.

Corrective actions are where many reviews collapse into abstraction. Statements such as “improve communications,” “strengthen training,” or “enhance coordination” sound responsible but do very little on their own. FEMA’s HSEEP doctrine and related improvement-planning templates are useful because they emphasize that corrective actions should be specific, monitored, and tied to actual capability improvement. In practical terms, that means the organization should define what must change, who owns the change, what timeline governs implementation, and how the organization will validate that the change actually improved readiness. A useful corrective action might involve validating executive briefing formats, testing identity reset sequencing, clarifying vendor escalation expectations, or exercising degraded operations under realistic staffing conditions. A vague corrective action produces vague improvement.

This chapter also matters because incidents and exercises are not the same, even though they should inform one another. FEMA's National Continuous Improvement Guidance notes that real-world incidents differ from exercises because they may last for extended periods, occur without warning, and create significant data-collection and resource challenges during analysis. That point is important for cyber defenders. A real incident usually generates more ambiguity, more emotional pressure, more incomplete records, and more institutional sensitivity than an exercise does. That does not make the review less important. It makes a disciplined review more important. Organizations should expect more friction in collecting facts, validating judgments, and analyzing communication or leadership performance after a real incident than they would after a controlled exercise.

For professionals moving from cyber warfighter to cyber defender roles, this chapter reinforces another important transition. The goal of the review is not to celebrate toughness, justify every action, or prove that the organization endured a difficult event. The goal is to determine whether the next incident will begin in a stronger environment because this one happened. Civilian critical infrastructure defense depends on institutional learning. If the organization cannot preserve the hard lessons of a real event and turn them into changes in architecture, identity discipline, leadership rhythm, communications structure, or continuity readiness, then it has not extracted the real value from the incident. It has merely survived it.

A useful standard for judging the review is therefore simple. If the next incident begins with the same avoidable weaknesses, then the lesson was documented but not learned. A review that changes decisions, structures, and readiness is an operational asset. A review that creates a tidy narrative without altering future performance is only paperwork.

Critical Thinking Question: What distinguishes an after-action review that genuinely improves readiness from one that merely documents what happened and allows the organization to move on without changing how it will respond next time?

Chapter 21: Building Long-Term Resilience

Recovery and resilience are related, but they are not the same thing. Recovery is the disciplined restoration of operations after an incident. Resilience is broader. It is the organization's enduring ability to absorb disruption, adapt under pressure, restore essential function, and improve before the next crisis arrives.[88] An organization can recover from a serious incident and still remain fundamentally fragile. If the same weaknesses in identity control, dependency awareness, leadership rhythm, communications, vendor reliance, and continuity planning remain largely unchanged, then recovery has restored function without materially improving resilience.

This distinction matters because the end of the incident often creates the beginning of drift. Once systems are restored, public attention fades, leadership pressure eases, and the visible crisis passes, the organization is tempted to treat completion as proof of improvement. That temptation is understandable. Incidents are exhausting. People want closure. Operations wants normalcy. Executives want reassurance that the worst is over. The problem is that resilience is not built in the moment when urgency is highest. It is built after urgency declines, when the organization decides which lessons are important enough to institutionalize.

NIST's guidance on cybersecurity event recovery supports this broader view by emphasizing planning, playbook development, testing, and continual improvement as part of recovery readiness.[88] That framing is important because it treats resilience as the product of repeated organizational decisions rather than as a generic aspiration. Resilience is not simply the ability to "bounce back." In critical infrastructure defense, it

---

[88] Michael Bartock, Jeffrey Cichonski, Murugiah Souppaya, Matthew Smith, Gregory Witte, and Karen Scarfone, Guide for Cybersecurity Event Recovery, NIST SP 800-184 (Gaithersburg, MD: National Institute of Standards and Technology, December 2016).

is the result of architecture, governance, training, trust management, operational realism, and disciplined follow-through over time.

Architecture is one of the most important foundations of that follow-through. Organizations become more resilient when they reduce unnecessary trust, improve segmentation, strengthen logging, clarify administrative pathways, document dependencies, and limit the ability of one compromised function to become a multi-system consequence event.[89] These changes are rarely glamorous, but they matter more than many visible security activities because they shape the environment the next incident will encounter. If the architecture remains flat in trust terms, weak in identity governance, or poorly understood in dependency terms, then the next incident begins with the same structural advantages the last one exploited.

This is one reason identity and trust must remain central after the incident is over. Earlier chapters emphasized that modern incidents often expand through accounts, administrative pathways, and trusted relationships rather than only through visible malware or single-system compromise. That lesson should not disappear once the environment is quieter. Resilience depends heavily on whether the organization knows which identities matter most, which privileges are truly necessary, which service accounts are over-trusted, which remote access paths are hard to govern, and which interdependencies turn a local problem into a wider continuity issue. If those trust conditions remain vague, resilience remains shallow.

The same principle applies to operational dependency. Critical infrastructure environments are rarely defined by isolated systems. They

---

[89] National Institute of Standards and Technology, The NIST Cybersecurity Framework (CSF) 2.0, NIST CSWP 29 (Gaithersburg, MD: National Institute of Standards and Technology, February 26, 2024); see also NIST SP 1303, NIST Cybersecurity Framework 2.0: Enterprise Risk Management Quick-Start Guide, October 21, 2024.

are defined by relationships among systems, services, vendors, support channels, manual workarounds, and continuity assumptions. CISA's dependency and resilience materials are useful here because they reinforce the reality that infrastructure consequences are often transmitted through dependencies rather than through direct compromise alone.[90] A resilient organization therefore does more than harden the system that failed last time. It improves its understanding of what depends on that system, what support paths keep it operating, what fallback modes exist, and how quickly the effects of disruption move across the broader environment.

Training and rehearsal are another major part of resilience, and this is where organizations often confuse awareness with capability. A team may know the policy, understand the response phases, and still perform poorly if it has not practiced under realistic friction. FEMA's continuous improvement approach and exercise doctrine are relevant here because they emphasize that organizations become stronger by repeatedly identifying issues, assigning corrective actions, and validating those improvements in practice rather than assuming that written lessons will naturally become operational habits.[91] In resilience terms, that means exercises should not merely confirm that a plan exists. They should test whether the organization has become better at recognizing weak signals, managing degraded operations, coordinating across teams, briefing leadership, and adapting when conditions are poor.

This is also why leadership and governance belong inside the resilience discussion rather than outside it. Many organizations speak

---

[90] Cybersecurity and Infrastructure Security Agency, "Critical Infrastructure Security and Resilience," accessed May 20, 2026; Cybersecurity and Infrastructure Security Agency, "Infrastructure Dependency Primer: Learn," accessed May 20, 2026.

[91] Federal Emergency Management Agency, "Continuous Improvement," September 14, 2023; Federal Emergency Management Agency PrepToolkit, "Continuous Improvement Beyond the AAR," accessed May 20, 2026.

about resilience as though it were primarily a technical property. It is not. Resilience depends on whether leaders preserve investment in controls and practices that reduce future consequences, whether governance processes keep attention on unresolved weaknesses after the crisis has faded, and whether the organization treats difficult lessons as obligations rather than optional improvements. NIST's CSF 2.0 and its enterprise risk management guidance are useful here because they position cybersecurity inside broader governance and risk processes rather than as a disconnected technical specialty.[92] That is exactly where resilience must live if it is going to survive leadership turnover, shifting priorities, and the gradual erosion of urgency.

Institutional memory is another essential part of resilience, and it is often underestimated. Organizations lose resilience when they simplify the incident into a cleaner story than the facts deserve. They say the issue was resolved, the systems were restored, and the organization performed well enough. Some of that may be true. But if the deeper lesson is not preserved in corrective actions, revised assumptions, updated priorities, and repeated practice, then the organization slowly returns to the same vulnerabilities with a stronger but misleading sense of confidence. FEMA's continuous improvement materials are useful on this point because they emphasize improvement as an ongoing capability, not just a post-event reporting exercise. The real threat to resilience is not only external attack. It is institutional forgetting.

For professionals moving from cyber warfighter to cyber defender roles, this final chapter should clarify the long view of the mission. In a civilian critical infrastructure context, success is not measured only by

---

[92] National Institute of Standards and Technology, The NIST Cybersecurity Framework (CSF) 2.0, NIST CSWP 29 (Gaithersburg, MD: National Institute of Standards and Technology, February 26, 2024); see also NIST SP 1303, NIST Cybersecurity Framework 2.0: Enterprise Risk Management Quick-Start Guide, October 21, 2024.

whether the defender performs well on the worst day. It is measured by whether the defender helps shape conditions so that the next bad day begins in a stronger environment. That means reducing trust sprawl, improving communication discipline, strengthening operational coordination, building realistic exercises, clarifying decision rights, and preserving difficult lessons after the crisis has passed. The defender's work therefore extends beyond response. It includes building the future operating conditions under which response will occur.

This is part of what makes critical infrastructure defense different from narrower cyber roles. The defender is not merely protecting data or systems in the abstract. The defender is helping protect functions people depend on, often without seeing the technical effort behind those functions. That responsibility requires more than technical skill. It requires patience, institutional seriousness, and a willingness to invest in improvements that may not become publicly visible until the next crisis reveals whether they were real.

In the end, resilience is not proven by slogans, maturity charts, or the organization's own confidence. It is proven by whether the organization becomes measurably harder to destabilize, easier to coordinate, and more disciplined in how it restores trust after disruption. If the same weaknesses remain in place, then the organization may have recovered, but it has not yet become resilient. If the incident leads to stronger architecture, clearer priorities, better identity governance, improved exercises, and more durable institutional learning, then the organization has done something far more valuable than returning to normal. It has raised the ground on which the next incident will begin.

That is the real long-term work of critical infrastructure cyber defense. It is not dramatic, and it is rarely finished. But it is the work that determines whether future disruption becomes a repeat of the last crisis or the beginning of a stronger response.

Critical Thinking Question: What changes must become permanent in an organization's architecture, governance, training, and institutional memory before it can honestly claim that it is more resilient after a major incident rather than merely recovered from one?

References

Cybersecurity and Infrastructure Security Agency, Federal Bureau of Investigation, Multi-State Information Sharing and Analysis Center, National Security Agency, U.S. Secret Service, & other partners. (2025). #StopRansomware guide. U.S. Department of Homeland Security. https://www.cisa.gov/stopransomware/ransomware-guide

Cybersecurity and Infrastructure Security Agency, Federal Bureau of Investigation, National Security Agency, & partner agencies. (2022, May 17). Weak security controls and practices routinely exploited for initial access. U.S. Department of Homeland Security. https://www.cisa.gov/news-events/alerts/2022/05/17/weak-security-controls-and-practices-routinely-exploited-initial-access

Cybersecurity and Infrastructure Security Agency, Federal Bureau of Investigation, National Security Agency, & partner agencies. (2024, June 18). Modern approaches to network access security. U.S. Department of Homeland Security. https://www.cisa.gov/resources-tools/resources/modern-approaches-network-access-security

Cybersecurity and Infrastructure Security Agency. (2016). National cyber incident response plan. U.S. Department of Homeland Security.

Cybersecurity and Infrastructure Security Agency. (2024). Federal government cybersecurity incident and vulnerability response playbooks. U.S. Department of Homeland Security. https://www.cisa.gov/resources-tools/resources/federal-government-cybersecurity-incident-and-vulnerability-response-playbooks

Cybersecurity and Infrastructure Security Agency. (2024, January 31). Incident response plan (IRP) basics. U.S. Department of Homeland

Security. https://www.cisa.gov/resources-tools/resources/incident-response-plan-irp-basics

Cybersecurity and Infrastructure Security Agency. (2025, July 30). Eviction strategies tool. U.S. Department of Homeland Security. https://www.cisa.gov/resources-tools/resources/eviction-strategies-tool

Cybersecurity and Infrastructure Security Agency. (2026, April 29). Adapting zero trust principles to operational technology. U.S. Department of Homeland Security.

Cybersecurity and Infrastructure Security Agency. (2026, April 29). CISA and U.S. government partners unveil guide to accelerate zero trust adoption in operational technology. U.S. Department of Homeland Security.

Cybersecurity and Infrastructure Security Agency. (n.d.). CISA tabletop exercise packages. U.S. Department of Homeland Security. Retrieved May 20, 2026, from https://www.cisa.gov/resources-tools/services/cisa-tabletop-exercise-packages

Cybersecurity and Infrastructure Security Agency. (n.d.). Critical infrastructure security and resilience. U.S. Department of Homeland Security. Retrieved May 20, 2026, from https://www.cisa.gov/topics/critical-infrastructure-security-and-resilience

Cybersecurity and Infrastructure Security Agency. (n.d.). Cybersecurity best practices for industrial control systems [PDF]. U.S. Department of Homeland Security.

Cybersecurity and Infrastructure Security Agency. (n.d.). ICS recommended practices. U.S. Department of Homeland Security.

Retrieved May 20, 2026, from https://www.cisa.gov/resources-tools/resources/ics-recommended-practices

Cybersecurity and Infrastructure Security Agency. (n.d.). ICS training available through CISA. U.S. Department of Homeland Security. Retrieved May 20, 2026, from https://www.cisa.gov/resources-tools/programs/ics-training-available-through-cisa

Cybersecurity and Infrastructure Security Agency. (n.d.). Incident response. U.S. Department of Homeland Security. Retrieved May 20, 2026, from https://www.cisa.gov/topics/cyber-threats-and-response/incident-response

Cybersecurity and Infrastructure Security Agency. (n.d.). Industrial control systems. U.S. Department of Homeland Security. Retrieved May 20, 2026, from https://www.cisa.gov/topics/industrial-control-systems

Cybersecurity and Infrastructure Security Agency. (n.d.). Infrastructure dependency primer: Learn. U.S. Department of Homeland Security. Retrieved May 20, 2026, from https://www.cisa.gov/topics/critical-infrastructure-security-and-resilience/resilience-services/infrastructure-dependency-primer/learn

Cybersecurity and Infrastructure Security Agency. (n.d.). Planning: Response & recovery. U.S. Department of Homeland Security. Retrieved May 20, 2026, from https://www.cisa.gov/planning-response-recovery

Cybersecurity and Infrastructure Security Agency. (n.d.). Sector risk management agencies. U.S. Department of Homeland Security. Retrieved May 20, 2026, from https://www.cisa.gov/topics/critical-infrastructure-security-and-resilience/critical-infrastructure-sectors/sector-risk-management-agencies

Cybersecurity and Infrastructure Security Agency. (n.d.). Use logging on business systems. U.S. Department of Homeland Security. Retrieved May 20, 2026, from https://www.cisa.gov/audiences/small-and-medium-businesses/secure-your-business/use-logging-on-business-systems

Cybersecurity and Infrastructure Security Agency. (n.d.). Use logging on government systems. U.S. Department of Homeland Security. Retrieved May 20, 2026, from https://www.cisa.gov/audiences/state-local-tribal-and-territorial-government/secure-us-sltt/use-logging-government-systems

Dempsey, K., Chawla, N., Johnson, L., Johnston, R., Jones, A., Orebaugh, A., Scholl, M., & Stine, K. (2011). Information security continuous monitoring (ISCM) for federal information systems and organizations (NIST SP 800-137). U.S. Department of Commerce. https://doi.org/10.6028/NIST.SP.800-137

Federal Emergency Management Agency. (2017). National incident management system (3rd ed.). U.S. Department of Homeland Security.

Federal Emergency Management Agency. (2023, September 14). Continuous improvement. U.S. Department of Homeland Security.

Federal Emergency Management Agency. (2025, January 27). Homeland Security Exercise and Evaluation Program. U.S. Department of Homeland Security. https://www.fema.gov/emergency-managers/national-preparedness/exercises/hseep

Federal Emergency Management Agency. (2025, July 28). National Incident Management System. U.S. Department of Homeland Security. https://www.fema.gov/emergency-managers/nims

Federal Emergency Management Agency. (n.d.). After action report. PrepToolkit National Continuous Improvement Guidance. Retrieved May 20, 2026, from https://preptoolkit.fema.gov/web/cip-citap/new-to-continuous-improvement/-/asset_publisher/NKZPnuTrRr0A/article/id/39547372

Federal Emergency Management Agency. (n.d.). Continuous improvement beyond the AAR. PrepToolkit National Continuous Improvement Guidance. Retrieved May 20, 2026, from https://preptoolkit.fema.gov/web/cip-citap/continuous-improvement-beyond-the-aar/-/asset_publisher/ujEgaXIbWQLW/article/id/39544098

Federal Emergency Management Agency. (n.d.). Improvement planning. PrepToolkit HSEEP Resources. Retrieved May 20, 2026, from https://preptoolkit.fema.gov/web/hseep-resources/improvement-planning

Federal Emergency Management Agency. (n.d.). NIMS components. U.S. Department of Homeland Security. Retrieved May 20, 2026, from https://www.fema.gov/emergency-managers/nims/components

Joint Task Force Transformation Initiative. (2012). Guide for conducting risk assessments (NIST SP 800-30 Rev. 1). U.S. Department of Commerce. https://doi.org/10.6028/NIST.SP.800-30r1

Kent, K., & Souppaya, M. (2006). Guide to computer security log management (NIST SP 800-92). U.S. Department of Commerce.

Kent, K., Chevalier, S., Grance, T., & Dang, H. (2006). Guide to integrating forensic techniques into incident response (NIST SP 800-86). U.S. Department of Commerce.

National Institute of Standards and Technology. (2006). Guide to test, training, and exercise programs for IT plans and capabilities (NIST SP 800-84). U.S. Department of Commerce.

National Institute of Standards and Technology. (2024). NIST Cybersecurity Framework 2.0: Enterprise risk management quick-start guide (NIST SP 1303). U.S. Department of Commerce.

National Institute of Standards and Technology. (2024). The NIST Cybersecurity Framework (CSF) 2.0 (NIST CSWP 29). U.S. Department of Commerce.

Nelson, A., Rekhi, S., Souppaya, M., & Scarfone, K. (2025). Incident response recommendations and considerations for cybersecurity risk management: A CSF 2.0 community profile (NIST SP 800-61 Rev. 3). U.S. Department of Commerce. https://doi.org/10.6028/NIST.SP.800-61r3

Petersen, R. C., Santos, D., Smith, M. C., Wetzel, K. A., & Witte, G. A. (2020). Workforce framework for cybersecurity (NICE Framework) (NIST SP 800-181 Rev. 1). U.S. Department of Commerce.

Rose, S., Borchert, O., Mitchell, S., & Connelly, S. (2020). Zero trust architecture (NIST SP 800-207). U.S. Department of Commerce.

Stouffer, K., Pease, M., Tang, C., Zimmerman, T., Pillitteri, V., Lightman, S., Hahn, A., Saravia, S., Sherule, A., & Thompson, M. (2023). Guide to operational technology (OT) security (NIST SP 800-82 Rev. 3). U.S. Department of Commerce.

Swanson, M., Bowen, P., Phillips, A. W., Gallup, D., & Lynes, D. (2010). Contingency planning guide for federal information systems (NIST SP 800-34 Rev. 1). U.S. Department of Commerce.

U.S. Department of Justice, Computer Crime and Intellectual Property Section. (n.d.). Reporting computer, internet-related, or intellectual property crime. Retrieved May 20, 2026, from https://www.justice.gov/criminal/criminal-ccips/reporting-computer-internet-related-or-intellectual-property-crime

U.S. Department of Justice. (n.d.). Justice Manual, 9-48.000 Computer Fraud and Abuse. Retrieved May 20, 2026, from https://www.justice.gov/jm/jm-9-48000-computer-fraud

U.S. Government Accountability Office. (2025, February 25). High-risk series (discussion of federal cybersecurity workforce management challenges and critical infrastructure entities).

U.S. Government Accountability Office. (2025, January 16). Cybersecurity workforce: Departments need to fully implement key practices (GAO-25-106795).

www.ingramcontent.com/pod-product-compliance
Lightning Source LLC
LaVergne TN
LVHW060547061126
851857LV00017B/1930